RADIATION PROTECTION FOR MEDICAL AND ALLIED HEALTH PERSONNEL

Recommendations of the
NATIONAL COUNCIL ON RADIATION
PROTECTION AND MEASUREMENTS

Issued October 30, 1989

National Council on Radiation Protection and Measurements
7910 WOODMONT AVENUE / Bethesda, MD 20814

Library of Congress Cataloging-in-Publication Data

National Council on Radiation Protection and Measurements.
  Radiation protection for medical and allied health personnel : recommendations of the National Council on Radiation Protection and Measurements.
    p. cm.—(NCRP report : no. 105)
    "Issued October 30, 1989."
    Supersedes NCRP report no. 48.
    Includes bibliographical references.
    ISBN 0-929600-09-6
    1. Hospitals—Radiological services—Safety measures.
2. Radiology, Medical—Safety measures. I. Title. II. Series.
    [DNLM: 1. Allied Health Personnel. 2. Radiation Injuries—prevention & control. 3. Radiation Protection—standards. WN 650 N277rb]
RA975.5.R3N37   1989
616.07'57'0289—dc20
DNLM/DLC
for Library of Congress   89-23872
                                                             CIP

Preface

The National Council on Radiation Protection and Measurements (NCRP) published Report No. 48, *Radiation Protection for Medical and Allied Health Personnel* in 1976. Many changes in medical practice and procedures involving ionizing radiation have occurred in the intervening 13 years. As a result, the Council determined to prepare this new report to supersede NCRP Report No. 48. The primary objective of this new report is to update the material to include new radiation sources used in medicine. In addition, an attempt has been made to reflect current practice in medicine and present the material in terms readily understood by an audience, most of whom have limited expertise in radiation protection terminology and principles. Although it is not designed as a guideline for the practicing health or medical physicist, it should be valuable in providing instruction and training of hospital personnel.

This report is intended to cover only those sources of ionizing radiation encountered commonly in the clinical environment. The less common types of radiation such as neutrons and pions are not discussed, principally because in those institutions where such sources are used, existing radiation safety programs should provide education and training to all of those needing it.

The first seven sections of this report provide general information on radiation and its uses in medicine for all readers. Section 8, Specific Guidelines, provides pertinent job related information for personnel involved with radiation sources. Each subsection of the specific guidelines was designed to stand alone. The length of each subsection is proportional to the potential for, or actual involvement with, radiation sources in a particular job category.

Providing specific guidance for every individual medical or paramedical specialty is beyond the scope of this report although personnel in all specialty groups should find this report helpful. For example, physicians, operating room nurses and respiratory therapists occasionally involved in x-ray procedures, will find information in Section 8 appropriate for their needs.

The International System of Units (SI) is used in this report followed by conventional units in parentheses in accordance with the procedure set forth in NCRP Report No. 82, SI Units in Radiation Protection and Measurements (NCRP, 1985a).

This report was prepared by Scientific Committee 46-6 on Radiation Protection for Medical and Allied Health Personnel which operated under the auspices of Scientific Committee 46 on Operational Radiation Safety.

Serving on Scientific Committee 46-6 were:

Kenneth L. Miller, *Chairman*
Pennsylvania State University
Hershey, Pennsylvania

David E. Cunningham
Pennsylvania State University
Hershey, Pennsylvania

Mary E. Moore
Cooper Hospital/
University Medical Center
Camden, New Jersey

L. Stephen Graham
Veterans Administration/UCLA
Sepulveda, California

Jean M. St. Germain
Memorial Sloan-Kettering
Cancer Center
New York, New York

Carol B. Jankowski
Brigham and Women's Hospital
Boston, Massachusetts

Scientific Committee 46 Liaison Member

William R. Hendee
American Medical Association
Chicago, Illinois

Serving on Scientific Committee 46 on Operational Radiation Safety were:

Charles B. Meinhold, *Chairman*
Brookhaven National Laboratory
Upton, New York

Ernest A. Belvin (1983-1987)
Tennessee Valley Authority
Chattanooga, Tennessee

Thomas D. Murphy
GPU Nuclear Corporation
Parsippany, New Jersey

William R. Casey (1983-1989)
Brookhaven National Laboratory
Upton, New York

David S. Myers (1987-)
Lawrence Livermore
   Laboratories
Livermore, California

Robert J. Catlin
Robert J. Catlin Corporation
Palo Alto, California

Keith Schiager
University of Utah
Salt Lake City, Utah

William R. Hendee
American Medical Association
Chicago, Illinois

Ralph H. Thomas (1989-)
Lawrence Berkeley
Laboratory
Berkeley, California

Kenneth R. Kase
University of Massachusetts
Worcester, Massachusetts

Robert G. Wissink
Minnesota Mining and
  Manufacturing Company
St. Paul, Minnesota

James E. McLaughlin
University of California
Los Angeles, California

Paul L. Ziemer
Purdue University
West Lafayette, Indiana

NCRP Secretariat

James A. Spahn, Jr. (1986-1989)
R. T. Wangemann (1986)
E. Ivan White (1983-1985)

The Council wishes to express its appreciation to the members of the Committee for the time and effort devoted to the preparation of this report.

Warren K. Sinclair
President, NCRP

Bethesda, Maryland
15 September 1989

Contents

1. General Considerations

1.1 Introduction

With the ever-increasing use of x rays and radioactive materials in medicine, more people may be exposed to ionizing radiations in the course of their work. The professional status of these individuals ranges from the highly-trained radiation specialist to the casual interdepartmental messenger. Many of these people have very little information about the possible biological effects of radiation, about the amounts which may be significant, or about ways to reduce their exposure to radiation. Their attitudes toward possible exposure vary from indifference to extreme fear. Frequently they have questions about radiation, radiation protection practices and the regulatory requirements but are reluctant or unable to seek out those who could provide answers. Their concern and interest, however, should not be ignored. This report seeks to meet their needs.

1.2 Purpose of Report

This report is intended to provide information about radiation, its effects on humans, protection against radiation and regulatory control requirements for those individuals who come into contact with radiation sources in the course of their work in medical facilities. It is aimed particularly at those individuals with limited training or experience in radiation matters. The goal is to provide easily understood information on radiation, its effects and radiation protection.

The report contains, in Section 8, material which will be of interest to the different categories of personnel working where radiation may be used. Administrators and supervisory personnel should find the report helpful in pointing out where possible hazards may exist. The report contains information about radiation protection for:

X-ray technologists and technicians and ultrasonographers
Nuclear medicine technologists and technicians
Nurses, aides, orderlies
Pathologists and Morticians

1

Non-Radiation Trained Physicians
Laboratory technicians
Shipping and receiving room personnel
Animal care personnel
Porters, janitors, maintenance personnel
Administrative Personnel
Engineering Personnel
In-house Fire Crews.

A copy of this report would be useful and should be made available to anyone desiring information about radiation protection because of concerns about radiation exposure as a result of their work.

1.3 Topics to be Considered

The radiation sources, uses, and facilities to be considered include the following:

X-Ray Diagnosis
    General Radiography
    Mobile (portable) Equipment
    Operating Room Procedures
    Special Radiographic Procedures
    Animal Radiography
Radiation Therapy
    X rays, Cobalt Teletherapy and Particle Accelerators
    Brachytherapy
        Sealed source storage area
        Patient and administration areas
        Post-administration care
Nuclear Medicine and Radioactive Materials (Radionuclides)
    The High Activity Laboratory
        Receiving area
        Work area
        Dose preparation area
        Dose administration area
        Storage
    Radiopharmaceutical Procedures
        Radioimmunoassay
        Bioassay
        In vitro testing
    Hospital Procedures

Therapeutic applications
Patient waiting area
Diagnostic tests
Research Laboratories
  Physics, chemistry, radiology, radiopharmaceuticals
Disposal Facilities for Solids, Liquids, Gases
  Hospitals
  Laboratories
Morgue
Animal housing rooms

Obviously, these topics cannot be covered in detail. Other reports from the National Council on Radiation Protection and Measurements (NCRP) provide details on some of the specified topics; the general purpose here is to point out where special precautions should be observed, and to prevent undue worry about situations which represent little risk.

In this report, unless stated otherwise, "radiation" means ionizing radiation, such as x rays, which is not to be confused with other forms of energy such as ultrasound. Information about these other forms is set out, however, in Section 8.15 and Appendix D.

The reader is referred to Appendix C for definitions of terms used in the report.

One point of terminology should be emphasized. In the various reports of the NCRP, the terms "*shall*" and "*should*" are used with strictly defined meanings: *Shall* indicates a recommendation that is necessary or essential to meet the currently accepted standards of protection. *Should* indicates an advisory recommendation that is to be applied when practicable. It is equivalent to "is recommended" or "is advisable". When these words occur in the text in such a manner as to refer to recommendations, they are italicized.

2. Radiation Exposure

The high quality of medical care that we have today would not exist without the use of radiation. Over the past 90 years, radiation has become an integral tool in the prevention, diagnosis and treatment of illness. Research laboratories use small quantities of radionuclides to learn more about normal body function and diseases and to develop better means of treating them. Diagnostic studies, such as dental x rays, lung scans, angiograms and computed tomographic (CT) scans all utilize ionizing radiation to demonstrate in detail the anatomic and physiologic features of sites of disease and injury in the body. Radiation therapy utilizes the cell-killing abilities of high-dose radiation to treat malignant conditions. Despite the benefits that radiation provides to health care, radiation exposure may pose some health risk to both patient and worker. An understanding of the sources of medically applied radiation and appropriate protective measures allows medical and other health personnel to work safely with or near sources of radiation.

Ionizing radiation may be emitted in a continuous manner by radioactive materials, both those used as medical sources and natural sources such as rocks and soil, or cosmic radiation from outer space. In general, the risk of exposure from radioactive materials continues until their radioactivity has been sufficiently diminished by radioactive decay processes. Radiation may also be produced by devices such as x-ray units or accelerators but only when the device is energized.

The types of non-ionizing radiation encountered in medical practice include ultrasound, radiofrequency radiation, which includes microwaves, and laser beams; these radiations are produced by energized devices and the non-ionizing radiation ceases when the device is switched off (See Appendix D).

2.1 Radiation Quantities and Units

Amounts of radiation and radioactivity are specified in terms of internationally accepted units. However, a transition in the units used is presently underway. All units in this report are expressed in

TABLE 2.1—*Frequently used SI prefixes*

Factor	Prefix	Symbol
10^{12}	tera	T
10^9	giga	G
10^6	mega	M
10^3	kilo	k
10^{-3}	milli	m
10^{-6}	micro	μ
10^{-9}	nano	n
10^{-12}	pico	p
10^{-15}	femto	f
10^{-18}	atto	a

terms of the international system of units, SI, with the corresponding value of the formerly employed unit following in parentheses. The absorbed dose received by humans from any source is expressed in units of gray (100 rad) and the dose equivalent in units of sievert (100 rem) or in multiples or submultiples of these units such as milligray (100 mrad) or millisievert (100 mrem) (see Table 2.1). The gray (100 rad) is used to express the absorbed dose in tissue; the sievert is used to express the dose equivalent which is a quantity in which the absorbed dose is weighted by the quality factor for the type of radiation which delivered the dose equivalent. Because x and gamma radiations are the reference radiations and their quality factor is 1, the numerical values of absorbed dose and dose equivalent are equal for these types of radiations, the ones most commonly used in medical applications.

[It has been established practice for many years to express the quantity of radiation in terms of the *exposure*, measured in *roentgens* (R). Exposure is a measure of the ionization caused by the absorption of x rays in a specified mass of air -- at the point of interest. In order to facilitate the use of the SI units, the quantity, *air kerma*, can be used for specification of irradiation. The unit of kerma is the gray (Gy). An exposure of 1 R corresponds to an air kerma of about 8.7 mGy.]

The frequency of radiation emissions from a radioactive material is related to the number of atoms transformed per second. Activity is the term used to specify the rate of spontaneous nuclear transformation of a radioactive nuclide. Becquerel (Bq) is replacing curie (Ci) as the unit of activity. An example of the use of these units is that typical injections for imaging purposes in nuclear medicine studies range in activity from 7.4 MBq (200 μCi) to 740 MBq (20 mCi).

2.2 Background Radiation

Many employees in medical facilities may be exposed on a daily basis to radiation from radioactive material or radiation-producing devices. Other employees may be exposed occasionally. Everyone, however, is exposed at all times to naturally occurring radiation sources in the environment. This radiation is referred to as natural background radiation and includes that from sources of cosmic and terrestrial origin as well as that from sources within the human body. Cosmic radiation penetrates and interacts with the earth's atmosphere thereby generating secondary radiation particles. The atmosphere absorbs some of this radiation, so that areas of higher elevation with less dense atmosphere receive more exposure from cosmic radiation than areas close to sea level. Similarly, passengers traveling in aircraft at 17 km (55,000 feet) are exposed to a higher dose equivalent rate (but for a shorter time), than passengers in conventional aircraft traveling at 11 km (35,000 feet). NCRP Report No. 94 (NCRP, 1988a) estimates that a transcontinental flight of 5 hours duration at 12 km (38,000 feet) results in a dose equivalent of 25 μSv (2.5 mrem) to the whole body.

The earth contains radioactive elements that have been present since the beginning of the planet itself. The intensity of terrestrial radiation varies by location, reflecting the different concentrations of radionuclides in the soil and underlying rock. Building materials, such as concrete and brick, may incorporate naturally occurring radioactive materials; exposure levels within buildings constructed of these materials are generally higher than the levels within wooden frame structures. Many buildings may have elevated levels of radon, a gaseous decay product arising from the decay of naturally occurring uranium-238 found in the soil. It has been estimated (NCRP, 1987d) that the average annual dose equivalent to the bronchial epithelium from radon decay products is approximately 24 mSv (2400 mrem or 2.4 rem).

Body tissues themselves are a source of natural radiation. Certain naturally-occurring radioactive atoms are taken into the body through ingestion and inhalation, and thereby accumulate in the tissues of the body, and contribute to the exposure of the individual. A significant component of the background dose equivalent to the body results from internally deposited potassium-40 (^{40}K), a component of food-stuffs and a very long-lived naturally occurring radionuclide. Table 2.2 provides a summary of average dose equivalent rates per year from natural background radiation sources in the United States.

In addition to natural background and radiation used for medical purposes, other sources of exposure to radiation can be found in the

TABLE 2.2—*Estimated total dose equivalent rate for a member of the population in the United States and Canada*[a] *from various sources of natural background radiation (mSv/y)*[b] *(from NCRP, 1988a).*

Source	Bronchial epithelium	Other soft tissues	Bone surfaces	Bone marrow
Cosmic	0.27	0.27	0.27	0.27
Cosmogenic[c]	0.01	0.01	0.01	0.03
Terrestrial	0.28	0.28	0.28	0.28
Inhaled	24.	—[d]	—[d]	—[d]
In the body[e]	0.35	0.35	1.1	0.50
Rounded totals	25.	0.9	1.7	1.1

[a]The dose equivalent rates for Canada are about 20% lower for the terrestrial and inhaled components.
[b]1 mSv = 100 mrem.
[c]Radionuclides produced when cosmic rays interact with atoms in the atmosphere, dose equivalent is primarily from Carbon-14 incorporated in tissues.
[d]Doses to other tissues from inhaled radionuclides included under "In the body."
[e]Excluding the cosmogenic component shown separately.

environment, although they contribute negligibly to average annual exposures. These include fallout from atomic weapons testing in the atmosphere, effluents from nuclear power plants and radioactivity in certain consumer products (*e.g.*, smoke detectors, tobacco products and radium-containing luminescent dial watches). (See NCRP Report No. 95 (1988b).

2.3 Patient Doses from Medical Sources

Other than natural background, the major source of radiation exposure to the U. S. population is that received by patients during the use of radiation in medicine and dentistry, primarily for diagnostic purposes [there were 1,240 diagnostic medical or dental procedures involving radiation exposure for every 1000 persons in the U.S. population in 1980 (NCRP, 1987d)]. Radiation from radiographic studies differs from background radiation in that exposure is normally restricted to a portion of the body and takes place over times that vary from a fraction of a second to minutes. Generally, radiation doses are calculated for the most radiosensitive organs. For example, a series of radiographs given for diagnosis of low back pain, or an upper GI series, provides a dose to the bone marrow of approximately 4 to 5 mGy (400 to 500 mrad) (FDA, 1977; NAS, 1980). A single chest film gives a much lower bone marrow dose, an average of 0.1 mGy (10 mrad) (FDA, 1977). Computerized tomography studies (CT scan) may provide an absorbed dose of more than 10 mGy (1,000 mrad) to the usually highly limited tissue volume subject to examination (Schonken *et al.*, 1978; Shope *et al.*, 1982). These partial body exposures can be taken into account by use of the effective dose equivalent (Report 91, NCRP, 1987a). The contribution of patient exposures in medical procedures to the annual effective dose equivalent of the U.S. population in terms of the average annual effective dose equivalent is 0.39 mSv (39 mrem) for diagnostic x rays, while that for nuclear medicine is 0.14 mSv (14 mrem). Thus the medical uses provide approximately 15 percent of the total average effective dose equivalent in the U.S. population (Report 100, NCRP, 1989a). Of course it needs to be recognized that in the case of patient exposure, the benefit of the medical procedure accrues directly to the individual exposed.

2.4 Medical Worker Exposures in the Medical Environment

Some employees (*e.g.*, physicians, radiological and nuclear medicine technologists) may be exposed to additional radiation above natural background because their occupation routinely requires working with or near sources of radiation. Most hospital employees, however, are not considered occupationally exposed workers and only occasionally come in contact with sources of radiation. For example, nurses may accompany a patient to the Nuclear Medicine Department and provide care following a diagnostic study. Operating room personnel frequently are present during fluoroscopic imaging of the operative site. Maintenance workers may be assigned to repair fume hoods or electrical wiring in a laboratory utilizing radionuclides. These situations generally will have been evaluated by the hospital Radiation Safety Officer (RSO) and can be expected to cause minimal exposure of workers when proper procedures are followed.

3. Biological Effects

3.1 Introduction

The discovery of x rays in 1895 and of radium in 1898 was followed rapidly by their application to human disease. However, it was soon evident that radiation could cause damage to tissues. Epilation (loss of hair), erythema (skin reddening) and other acute somatic effects of radiation exposure were the first symptoms noted in patients as well as in those physicians and physicists who first worked with radiation sources. (The exposures in those days were commonly hundreds of times greater than the ones typically received today.) Investigators irradiated living organisms in an attempt to understand the mechanisms responsible for the biological effects of radiation. It was found that certain tissues or organisms were more sensitive to radiation than others, particularly if their cells were rapidly dividing, such as is the case for cells of the hematopoietic system.

Following World War II, studies were initiated to investigate the effects of radiation on the Japanese populations who survived the atomic bombing of Hiroshima and Nagasaki. These studies are continuing today. The results of health studies of other groups and the results of A-bomb survivor studies are compared for consistency between findings. These groups include individuals who received exposure to radiation in their occupations as well as patients who were treated with radiation for a variety of conditions and diseases. The reports of the United Nations Scientific Committee on the Effects of Atomic Radiation (UNSCEAR) and the National Academy of Sciences Committee on the Biological Effects of Ionizing Radiation (BEIR) are comprehensive reviews of most of these data (UNSCEAR, 1986; 1988; NAS, 1980; 1988).

Numerous radiobiological studies have been conducted in animals, (e.g., mouse, rat, hamster, dog), and in cells and tissue cultures. Extrapolations to human beings from these experiments are problematic and despite the large amount of data accumulated, uncertainties remain regarding the effects of radiation at low doses and low dose rates. The most reliably estimated risks are those associated with doses of 1 Gy (100 rad) or more. There is general agreement that risks at smaller doses are at least proportionally smaller (e.g.,

9

no more than 1/10 the risk at 1/10 the dose), but it seems likely that they may, in fact, be considerably smaller (NCRP, 1980a).

Because the risk is small and because of the possibility of other competing nonradiation causes, it is difficult to observe radiation effects at low dose levels. Theoretically, it might be possible to study large populations for long periods of time but the number required, and the level of control necessary to rule out confounding from other causes of variation in human effects make investigations on such a scale impracticable (Land, 1980).

The serious radiation-induced diseases of concern in radiation protection fall into two general categories: stochastic effects and nonstochastic effects.

For the purposes of this Report, a stochastic effect is defined as one in which the probability of occurrence increases with increasing absorbed dose but the severity in affected individuals does not depend on the magnitude of the absorbed dose. A stochastic effect is an all-or-none response as far as individuals are concerned. A stochastic effect might arise as a result of radiation injury of a single cell or substructure such as a gene and is assumed to have no absolute dose threshold, despite the fact that currently available observations in population samples do not exclude zero effects at low radiation levels. Cancers (solid malignant tumors and leukemia) and genetic effects are regarded as the main stochastic effects or risks to health from exposure to ionizing radiation at low absorbed doses (NCRP, 1987a).

A nonstochastic effect of radiation exposure is defined as a somatic effect which increases in severity with increasing absorbed dose in affected individuals, owing to damage to increasing numbers of cells and tissues. Nonstochastic late effects, *e.g.*, diseases characterized by organ atrophy and fibrosis, are basically degenerative, as contrasted with the neoplastic growth characteristic of cancer. In general, considerably larger absorbed doses are required to cause nonstochastic effects to a degree of severity which seriously impairs health, as compared with absorbed doses required for a significant increase in cancer incidence. The incidence of nonstochastic effects in a population may increase with increasing absorbed dose, owing to differences in susceptibility and other contributing causes among individuals in the population. Examples of nonstochastic effects attributable to radiation exposure are lens opacification, blood changes, and a decrease in sperm production in the male. (NCRP, 1987a).

3.2 Acute Radiation Effects

Acute radiation effects (erythema, epilation, nausea, diarrhea) are those that appear within a short enough period of time after exposure to make it obvious that radiation was the cause. Acute effects have been observed only following high dose exposures, typically greater than 1 Gy (100 rads) to the whole body. The severity of the acute radiation effects observed following high doses is dependent upon the amount of tissue exposed, the nature of the tissue exposed, the dose rate and the total dose received. The potential for exposures that would result in acute effects generally does not exist in medical facilities.

3.3 Cancer

The most serious delayed effect of radiation is cancer. Radiation induced cancers arise years or decades after exposure and they are indistinguishable from those, much more frequent ones, that are due to other causes. These characteristics make it difficult to provide firm numerical estimates but it has been generally agreed that the general risk of developing cancer in a lifetime, which is 33 percent (SEER, 1981), is increased by about one percent by a whole body dose of 100 mGy (10 rad) (UNSCEAR, 1988).

While an increase of cancer incidence was noted in some of the early radiation workers, current exposure levels are so low that the excess incidence in radiation workers, although probably not zero, is statistically undetectable. The average exposure to medical personnel in the U.S. is below 10 mGy (1 rad) per year and it can be calculated that the increased risk of dying of cancer because of continuing exposure even at limits permissible during a working life may be of the order of 1 percent. This is similar to the figures in other "safe industries" which have a fatality risk of 1 or 2 percent. Of course, because of careful radiation protection practices, no workers are continuously exposed at the permissible limits.

3.4 Genetic Effects

A genetic effect of radiation is one that is transmitted to the offspring of the exposed individual. Radiation can impart energy to the germ cell nucleus, thereby causing breakage or alteration of molecular bonds which can result in mutation or chromosome breakage.

Radiation induced mutations do not differ from spontaneously induced mutations. At exposures typically received in today's medical setting, the probability of radiation-induced genetic effects is very small. Even in the case of the Japanese A-bomb survivors, who were exposed at higher levels, no significant excess of genetic effects has been observable.

3.5 Embryonic and Fetal Effects

The embryo or fetus is comprised of large numbers of rapidly dividing and radiosensitive cells. The amount and type of damage which may be induced are functions of the stage of development at which the fetus is irradiated and the absorbed dose.

Radiation received during the pre-implantation period can result in spontaneous abortion or resorption of the conceptus. Radiation injury during the period of organogenesis (2 to 8 weeks) can result in developmental abnormalities. The type of abnormality will depend on the organ system under development when the radiation is delivered. Radiation to the fetus between 8 and 15 weeks after conception increases the risk of mental retardation (Otake and Schull, 1984) and has more general adverse impact on intelligence and other neurological functions. The risk decreases during the subsequent period of fetal growth and development and, during the third trimester, is no greater than that of adults.

Special limits have been established for occupationally exposed pregnant women to ensure that the probability of birth defects is negligible.

4. Dose Limits

4.1 Dose Limits for Radiation Workers and Others

Occupational and public dose equivalent limits have been recommended by the NCRP (Table 4.1). These limits do not include exposure from natural background and exposures received as a patient for medical purposes. Occupationally exposed workers are limited to an annual effective dose equivalent of 50 millisievert (5000 millirem); the dose equivalent limits recommended for the general public generally are one-tenth or less of those for occupationally exposed individuals (NCRP,1987a). Partial body exposures and exposures of individual organs are accounted for by establishing the limits in terms of the effective dose equivalent, which weights the dose equivalent in terms of the risks resulting from partial body or organ exposure. Students under the age of 18 who are training in jobs with a potential for exposure should not receive more than 1 mSv (100 mrem) per year from their educational activities.

Some organs and areas of the body are less sensitive to radiation than others. As a result, for nonstochastic effects, the recommended annual occupational dose equivalent limit to the lens of the eye is 150 mSv (15,000 mrem); the annual dose equivalent limit recommended for other organs is 500 mSv (50,000 mrem).

4.2 Dose Limits for the Embryo and Fetus

The occupational exposure of pregnant or potentially pregnant women is an area of special concern (See Section 3.5). NCRP Report No. 53 (NCRP, 1977a) has specifically addressed this subject, and Report No. 91 (NCRP, 1987a) has given it further consideration, recommending special limits for the embryo/fetus. Although the mother can be considered as an occupationally exposed individual, the fetus cannot. Any exposure of the abdomen of a pregnant woman may also involve exposure of the fetus. The use of a surface dose as an estimate of the dose to the fetus fails to consider the attenuation of radiation in overlying tissue and amniotic fluid. Use of surface doses, therefore, will normally overestimate the fetal dose. Internal

13

TABLE 4.1—*Summary of recommendations*[a] (After Report No. 91, NCRP, 1987a)

A. Occupational exposures (annual)[b]
1. Effective dose equivalent limit   50 mSv   (5 rem)
    (stochastic effects)
2. Dose equivalent limits for tissues and
    organs (nonstochastic effects)
    a. Lens of eye   150 mSv   (15 rem)
    b. All others (e.g., red bone marrow,   500 mSv   (50 rem)
    breast, lung, gonads, skin and
    extremities)
3. Guidance: Cumulative exposure   10 mSv x age   (1 rem x age in years)

B. Public exposures (annual)
1. Effective dose equivalent limit,   1 mSv   (0.1 rem)
    continuous or frequent exposure[b]
2. Effective dose equivalent limit,   5 mSv   (0.5 rem)
    infrequent exposure[b]
3. Remedial action recommended when:
    a. Effective dose equivalent[c]   >5 mSv   (>0.5 rem)
    b. Exposure to radon and its decay >0.007 Jhm^{-3} (>2 WLM)
    products
4. Dose equivalent limits for lens of eye,   50 mSv   (5 rem)
    skin, and extremities[b]

C. Education and training exposures (annual)[c]
1. Effective dose equivalent limit   1 mSv   (0.1 rem)
2. Dose equivalent limit for lens of eye,   50 mSv   (5 rem)
    skin and extremities

D. Embryo-fetus exposures[b]
1. Total dose equivalent limit   5 mSv   (0.5 rem)
2. Dose equivalent limit in a month   0.5 mSv   (0.05 rem)

E. Negligible Individual Risk Level (annual)[b]
1. Effective dose equivalent per source or   0.01 mSv   (0.001 rem)
    practice

[a]Excluding medical exposures.
[b]Sum of external and internal exposures.
[c]Including background but excluding internal exposures.

dose from certain ingested or inhaled radionuclides may represent a particular hazard if such materials can cross the placenta and be incorporated into fetal tissue.

Premenopausal female radiation workers *shall* be informed of the risks to which the fetus may be exposed and the methods available for reducing exposure. Individual counseling for these women *should* be available. Included in any evaluation of risk and exposure will be existing personnel monitoring records, surveys of the workplace and a review of the sources of radiation. If this evaluation indicates the possibility of a dose equivalent to the fetus in excess of 5 mSv (500 mrem) during the gestation period, the employee *should* discuss her options with her employer. Once a pregnancy is made known by the employee, exposure of the embryo-fetus *should* be no greater than 0.5 mSv (50 mrem) in any one month.

4.3 Annual Occupational Doses

Average annual occupational whole body dose equivalents to medical personnel who are monitored for radiation exposure have been compared with those from other types of employment (Table 4.2). The mean dose equivalent to medical personnel who work with x rays or radiopharmaceuticals averages 1.0 to 1.4 mSv (100 to 140 mrem); similarly categorized dental personnel average 0.2 mSv (20 mrem). Annual dose equivalents for industrial workers are similar; monitored nuclear power plant employees average 5.6 mSv (560 mrem), while, for industrial radiographers, the mean dose equivalent is 2.8 mSv (280 mrem). All of these occupational doses are well below the limits, presumably because radiation safety personnel and radiation workers conscientiously follow good protection practices, and strive to keep doses as low as reasonably achievable.

TABLE 4.2—*Comparison of mean annual dose equivalents and collective dose equivalents for monitored workers (From NCRP, 1989b).*

Occupation	Number of workers (thousands)	Mean dose equivalent (mSv)[a]	Collective dose equivalent (person-SV)[b]	Year
Dentistry	259	0.2	60	1980
Private medical practice	155	1.0	160	1980
Hospital	126	1.4	170	1980
Industrial radiography	8.5	2.8	24	1985
Nuclear power plant worker	98	5.6	552	1984

[a] 1 mSv = 100 mrem
[b] 1 person-SV = 100 person-rem

4.4 Radiation Protection Philosophy: ALARA

The general philosophy followed by most institutions in minimizing radiation dose is that all exposures must be justified and, further, that they must be kept as low as reasonably achievable (ALARA), economic and social factors being taken into account. The ALARA concept applies to radiation workers as well as to the general public. The ALARA statement represents a commitment on the part of the institution to provide the resources and environment in which ALARA can be implemented. The NCRP recommends continuing efforts to maintain personnel exposures below allowable limits and to keep exposures as low as reasonably achievable.

An important part of an ALARA program is an annual administrative review of working conditions and personnel monitoring records. In this review, the roles of the RSO and the Radiation Safety Committee (RSC) in the implemention of goals as defined by the radiation protection program are examined.

The ALARA approach to radiation exposure management requires that the workers be aware of the rules governing the work situation. A training program, which informs the workers of any hazards in the work environment and methods to minimize these hazards, is essential to this approach (NCRP 1978a; 1983a).

5. Management of a Radiation Protection Program

5.1 Introduction

Because there is concern that there may be risks from low doses of ionizing radiation, it is prudent to make every effort to keep such exposures as low as reasonably achievable. An effective radiation protection program requires a commitment to radiation safety by everyone, including the management and all employees, not just radiation workers and radiation safety personnel.

5.2 Guidelines and Regulations

Radiation has been studied extensively, and guidelines and regulations dealing with all aspects of radiation safety and all types of radiation-producing sources have been developed by state and federal agencies, for the most part based upon recommendations of various radiation protection advisory groups [e.g., the National Council on Radiation Protection and Measurements (NCRP), and the International Commission on Radiological Protection (ICRP)].

Individuals or institutions wishing to possess radioactive materials in other than exempt amounts are required to obtain licenses issued by either the U.S. Nuclear Regulatory Commission or an equivalent agency at the state level. The issuance of these licenses is preceded by a complete review of the applicant's radiation protection program to ensure that it is adequate. A license for possession of radioactive materials carries with it the responsibility of ensuring that these materials will be handled, used and, ultimately, disposed of in a safe manner. Individuals or institutions holding such licenses are subject to periodic audits by licensing agencies. If, during these audits, significant deviations from either the license conditions or the routinely accepted safety practices are detected, the licensee is penalized commensurate with the potential hazard detected. Such penalties can be in the form of written notices of violations, fines or other penalties, including revocation of the license and immediate cessation of all activities involving the radiation sources. Most states place similar requirements on the use of other radiation sources such as x-ray machines and, in a few states, particle accelerators.

The management of each institution is responsible for ensuring that all license conditions, regulations and appropriate safety precautions are followed rigidly. In order to meet the requirements of the license, a formal radiation safety structure must be in place, including a Radiation Safety Committee (not required for every license) and a Radiation Safety Officer.

5.3 Radiation Safety Committee (RSC) and Radiation Safety Officer (RSO)

The size of the program and federal or state licensing requirements will determine the size of and the need for a radiation safety committee. The RSC's primary responsibility is to develop and maintain an effective radiation safety program for the medical facility (see also Section 8.1.2). To do this, its members must possess adequate knowledge of the principles of radiation physics and radiation protection. The membership of the Committee should include such individuals as a nuclear medicine physician, a radiologist, a radiation oncologist, a senior hospital administrator, a health or medical physicist, a senior nurse, an internist, and an investigator who uses radiation in research activities.

The RSO *should* be an individual with extensive training and education in areas such as radiation protection, radiation physics, radiation biology, instrumentation, dosimetry and shielding design. The designated RSO *should* be a health or medical physicist, but may be a physician or other individual qualified by virtue of experience or training. The primary function of the RSO is the supervision of the daily operation of a radiation safety program to ensure that individuals are protected from radiation. To do this, the RSO *should* report directly to top management and have ready access to all levels of the organization. NCRP Report No. 59 (NCRP, 1978a) describes these administrative arrangements in detail.

5.4 Records

Records dealing with all aspects of the radiation protection program are important to ensure compliance with regulations and license conditions, to provide internal review capabilities and to reduce liability. Periodic review of these records can identify trends that require corrective action as well as deficiencies in the radiation safety program. Such records include reports on contamination and radiation surveys, results of personnel monitoring, instrument and equipment calibrations, and documentation of training programs for employees. Records must also be maintained that document immediate and appropriate response to accidents, such as contaminations, spills, over-exposures, etc., as well as the corrective action taken to prevent similar occurrences.

5.5 Training and Continuing Education

Changes that occur in instrumentation, monitoring methods, recommendations and regulations make it imperative that all individuals involved in the use of ionizing radiation sources receive initial and continuing training and education. Such training can range from informal interdepartmental reviews to structured and accredited continuing education programs. It is the responsibility of management as well as of radiation workers to maintain a professional level of training and expertise. Management *should*, therefore, provide radiation workers with the opportunity to attend training and continuing education programs (NCRP, 1983a).

5.6 Personnel Monitoring

Personnel monitoring is recommended for individuals for whom there is a reasonable probability of exceeding 25 percent of the occupational dose equivalent limit of 50 mSv/y (5 rem/y) in the course of their work (NCRP, 1978a). In the medical environment, the majority of personnel occupational radiation exposures are below the level at which personnel monitoring is required. Nevertheless, most hospital personnel who work with radiation wear a personnel monitoring device [*e.g.*, film badge or thermoluminescent dosimeter (TLD)] to assess actual exposure during work or as a check against unplanned exposures. In some situations, workers may be asked to wear a dosimeter (film badge, TLD or pocket ionization chamber) only during a time of potential exposure (*e.g.*, operating room nurses assisting with surgical implantation of radioactive sources).

6. Sources of Radiation Exposure in the Medical Environment

In the medical environment, radiation exposure can arise either from materials (radionuclides) that spontaneously produce radiation or from devices that produce x rays or particulate radiation such as high energy electrons.

6.1 Radioactive Materials

For an understanding of the safety procedures relating to radioactive materials, some knowledge of the phenomenon of radioactivity is essential. The naturally occurring isotopes of most elements have stable nuclei, but some natural and many man-made elements have atoms whose nuclei are unstable and will eventually undergo radioactive transformation. Atoms with unstable nuclei spontaneously transform (decay) and release energy. Following this process they ultimately reach a stable state. The energy released may be in the form of a high-speed particle (beta or alpha) ejected from the nucleus, an electromagnetic wave (x-ray or gamma), or a combination of these. Fifty percent of all the atoms of a given radionuclide will transform during its characteristic time period called the half-life. During each succeeding half-life, 50 percent of the remaining atoms will be transformed, and after ten half lives the number of radioactive atoms has decreased to less than 1/10 of 1 percent. Half-lives of radionuclides vary from a fraction of a second to billions of years. Those that are used for medical purposes, either for diagnosis or therapy, have half-lives ranging from a few minutes to many years; generally short-lived radionuclides are used.

There are several types of radiation that can be emitted from radioactive atoms. From the standpoint of radiation protection, or clinical applications, it is important to be familiar with the nature of the radiation. The basic types of radiation are: alpha particles, negative and positive beta particles, characteristic x rays, gamma

20

rays and heavy particles from spontaneous fission. Some radionuclides emit just one of these types, others emit two or three.

Alpha radiation is easily absorbed and it will not penetrate the walls of common containers. It can also be stopped by a few centimeters of air. Radionuclides which emit alpha radiation are rarely used in medicine.

Beta radiations (electrons or positrons) are high speed electrons ejected from a nucleus during transformation. Such particles can pass through thin-walled containers. High-energy beta particles can penetrate a few millimeters into living tissues. Positive beta particles are called positrons, and they are always accompanied by high energy electromagnetic radiation (photons).

Gamma radiation is emitted from nuclei and it has electromagnetic properties that are identical to those of x rays. Gamma rays have a wide range of energies and penetrating abilities. Some radionuclides that are used in medicine emit both gamma rays and beta particles.

The previous discussions on penetrating ability of the various types of radiation are of limited consequence for internal emitters; *i.e.* radionuclides that have been taken into the body.

A list of radionuclides currently used in research, diagnostic or therapeutic procedures is provided in Table 6.1, with data concerning the half-lives and radiations emitted. See also Report No. 70 (NCRP, 1982) and Report No. 58 2nd edition (NCRP, 1985b).

6.1.1 *Unsealed Sources*

Unsealed sources of radionuclides may be found in several locations within a hospital. They are used in the clinical laboratory for analyzing blood samples, in the research laboratory for *in vitro* and animal studies, and in the nuclear medicine department for both diagnosis and therapy.

Diagnosis

In nuclear medicine, agents specifically targeted to an organ or organ system are labeled with radionuclides (these labeled agents are usually called radiopharmaceuticals) and administered to the patient.

Most diagnostic procedures in nuclear medicine involve direct measurement of the amount or distribution of radioactive material within the patient using an instrument known as a scintillation

TABLE 6.1—*Selected radionuclides used in medicine*

Radionuclide	Symbol	Half-life	Radiation
Hydrogen-3	^{3}H	12.3 y	Beta −
Carbon-11	^{11}C	20 min	Beta +
Nitrogen-13	^{13}N	10 min	Beta +
Carbon-14	^{14}C	5730 y	Beta −
Oxygen-15	^{15}O	2 min	Beta +
Fluorine-18	^{18}F	110 min	Beta +
Sodium-22	^{22}Na	2.6 y	Beta + ; Gamma
Sodium-24	^{24}Na	15 h	Beta − ; Gamma
Phosphorus-32	^{32}P	14 d	Beta − ; x rays (Bremsstrahlung)
Sulfur-35	^{35}S	87 d	Beta −
Chromium-51	^{51}Cr	28 d	Gamma
Cobalt-57	^{57}Co	271.8 d	Gamma
Cobalt-58	^{58}Co	71 d	Beta + ; Gamma
Iron-59	^{59}Fe	45 d	Beta − ; Gamma
Cobalt-60	^{60}Co	5.3 y	Beta − ; Gamma
Copper-64	^{64}Cu	12.7 h	Beta + ; Beta − ; Gamma
Gallium-67	^{67}Ga	78 h	Gamma
Gallium-68	^{68}Ga	68 min	Beta + ; Gamma
Selenium-75	^{75}Se	120 d	Gamma
Krypton-81m	^{81m}Kr	13 s	Gamma
Rubidium-81	^{81}Rb	4.6 h	Beta + ; Gamma
Molybdenum-99	^{99}Mo	66 h	Beta − ; Gamma
Technetium-99m	^{99m}Tc	6 h	Gamma
Indium-111	^{111}In	2.8 d	Gamma
Indium-113m	^{113m}In	99 min	Gamma
Tin-113	^{113}Sn	115 d	Gamma
Iodine-123	^{123}I	13 h	Gamma
Iodine-125	^{125}I	60 d	X Ray; Gamma
Xenon-127	^{127}Xe	36 d	Gamma
Iodine-131	^{131}I	8 d	Beta − ; Gamma
Xenon-133	^{133}Xe	5.2 d	Beta − ; Gamma
Barium-133	^{133}Ba	10.5 y	Gamma
Cesium-137	^{137}Cs	30 y	Beta − ; Gamma
Ytterbium-169	^{169}Yb	32 d	Gamma
Gold-195m	^{195m}Au	30 s	Gamma
Gold-195	^{195}Au	183 d	Gamma
Thallium-201	^{201}Th	3 d	X Ray; Gamma
Americium-241	^{241}Am	432 y	Alpha; Gamma
Radium-226	^{226}Ra	1602 y	Alpha; Gamma

camera. These measurements are made directly on the patient and require the administration of a radiopharmaceutical in tracer amounts. Such tests may be called uptakes, scans, imaging procedures or dynamic function studies.

The use of radioactivity in the analysis of body fluids does not always involve administering radioactive materials to the patient. In some tests, such as the radioimmunoassay (RIA) test, small amounts of radioactive materials are added to fluids or specimens extracted from the patient. In these tests, there is no radiation dose absorbed by the patient.

Therapy

Unsealed sources of radionuclides for therapeutic uses may be found in the departments of nuclear medicine, endocrinology or radiation therapy. Relatively large doses of these radionuclides (such as ^{131}I and ^{32}P) are used for treatment of various pathological conditions. During such a procedure the patient may be a significant source of radiation exposure for attending personnel, family and visitors. For further information see Sections 8.8, 8.9, 8.12 and Appendix B.

6.1.2   *Sealed Sources*

Certain cancers are treated with sealed radioactive sources placed directly in tissue, in a body cavity, or on body surfaces. This type of treatment is known as brachytherapy. Four basic types of implants can be found in the medical environment: temporary intracavitary implants; temporary interstitial implants; permanent interstitial implants; and plaques or applicators. Patients undergoing brachytherapy are usually confined to controlled areas within the hospital.

Sealed sources may also be used in certain instruments such as bone densitometers, gas chromatographs and blood irradiators. In nuclear medicine, sealed sources of ^{137}Cs, ^{60}Co, ^{57}Co, ^{133}Ba, or ^{241}Am are used for quality control of imaging instruments and radionuclide dose calibrators. Low activity sources of these same radionuclides are also used for testing and checking counting instruments. Sealed sources of high activity are used in external beam radiation therapy (see section 6.2.2). Unless specifically exempted, all sealed sources, including teletherapy sources discussed in section 6.2.2, are required to be leak tested at periodic intervals (usually 6 months maximum) to ensure detection of inadvertent escape of the radioactive material.

6.1.3   *Research*

Radioactive materials are used in virtually every phase of biomedical research. Many medical institutions have one or more clinical research laboratories using radioactive tracers. A detailed discussion on guidelines for personnel working in clinical and research laboratories is found in Section 8.3.

6.2 Radiation-Producing Equipment

6.2.1 *Diagnostic*

X rays are produced when high speed electrons are directed onto a metal target contained in a sealed glass vacuum tube called an x-ray tube. When the electrons strike the target, they produce a large amount of heat and lose some of their energy in the form of x rays which emerge in all directions. The x-ray tube is inserted into a specially designed lead container called the "housing". The housing is designed to limit the exit of x rays to one designated area called the port or window. The result is a beam, called the useful beam, which is directed at the region of interest in the body.

Because of the design of diagnostic x-ray units, x rays are produced only when the exposure switch is engaged. As soon as the switch is released, or the pre-set exposure time is reached, x-ray production ends. There are no residual x rays in the room, and the patient does not become radioactive.

In hospitals, diagnostic x-ray studies (radiographic and fluoroscopic) are performed primarily in the radiology department. Such studies are also performed in other areas, *e.g.*, operating rooms, intensive care units, coronary care units, special care nurseries, cardiac catheterization laboratories, emergency departments, and patients' rooms. Since the use of x rays in hospitals is widespread, the hospital staff should be familiar with and follow the basic radiation safety precautions and practices described in Section 7.

6.2.2 *Therapeutic*

In a radiation therapy department patients receive treatment for cancer and certain benign conditions. The type and energy of the radiation used depend on the type and location of the cancer. The most common types of therapy equipment are cobalt teletherapy equipment and linear accelerators. The patient does not become radioactive as a result of cobalt teletherapy. Linear accelerators, when operated above 10 MeV, are capable of producing neutrons in addition to high energy electromagnetic radiation. Patients who have been so treated may contain some radionuclides which emit radioactivity as they decay, but the half-lives of these radionuclides are short and dose rates from the surface of the patient do not represent a significant hazard to personnel in close proximity to the patient (Report No. 79, NCRP, 1984).

Cobalt-60 Teletherapy Units

Cobalt-60 (^{60}Co) teletherapy units use the gamma rays emitted from a sealed source. Occasionally other radionuclides such as cesium-137 are employed. The units are designed to provide continuous shielding of the radioactive source unless a treatment is in progress. When treatment starts, the ^{60}Co source is remotely positioned to allow the radiation beam to be aimed at the tumor. Upon completion of the treatment, the source is returned to its shielded position. No radioactivity is created in the patient.

Linear Accelerators

Linear accelerators produce high-energy x rays and/or high-energy electron beams. Like x-ray machines, these units do not produce a beam unless energized; therefore, there is minimal risk of radiation exposure to personnel or patients while positioning the patient for treatment prior to turning the machine on. One of the design advantages of high-energy linear accelerators is that the equipment can generate x rays or electrons. The electrons can be used for treatments by removing the target used to produce x rays and aiming the electron beam directly at the tumor. Because of the limited penetration of electrons in the body, electron beam therapy is used primarily for shallow depth tumors (*e.g.*, those of the head and neck, chest wall, skin).

Other Equipment

A radiation therapy department may also have lower energy x-ray equipment such as orthovoltage and/or superficial units. These units are not as prevalent as they were in the past because electron beam therapy has become available.

6.2.3 *Use of Radiation-Producing Equipment for Research*

Institutions with large biomedical research programs may have radiation-producing equipment for research, including analytical equipment (*e.g.*, x-ray diffraction units), cyclotrons for isotope production, or experimental treatment units (*e.g.*, neutron therapy facilities). This equipment should be used in a well-controlled environment. Employees should be aware of the safety systems present in the installation and training *shall* be provided in the use of these systems. The potential for exposure may be high, and radiation protection programs for these installations *shall* include an evaluation of the working environment, provision for fail-safe access control and interlock systems and special shielding where applicable. Negligence on the part of employees or the deliberate violation of safety systems should be reported immediately to the RSO and the management, not only because of the potential problems for the employee's own health, but also because of the safety problems that may result for other employees and visitors.

6.3 Other Radiation Sources

There frequently are sources of other types of radiation in medical facilities such as ultrasound generators, magnetic resonance equipment, and laser devices. Appendix D provides a brief discussion of some of these other types of equipment.

7. Basic Principles of Radiation Protection

7.1 Introduction

Radiation-producing equipment and radioactive materials provide significant diagnostic and therapeutic benefits to the patient. However, if used improperly or when accidents occur during use, they can present a risk to both patient and user. It is imperative that good radiation safety practices be used at all times. This section reviews the basic principles of radiation protection. Specific applications of these principles are incorporated into the specialty discussions in Section 8.

7.2 Control of External Exposure

Protection against radiation from either devices or radioactive material requires an understanding of the particular characteristics of the radiation involved. Measurement and identification of both the source and type of radiation involved are necessary before prescribing safety precautions. Once this is done, the RSO, in consultation with others, can formulate specific protective procedures for everyone involved. There are certain fundamental principles of radiation protection that should be understood by anyone who might be exposed to radiation. These involve the protection that can be provided by time, distance and shielding. These are discussed below.

7.2.1 *Time*

When exposed to radiation at a constant rate, the total dose equivalent received depends on the length of time exposed. A typical chest x ray exposes the patient to a high dose rate; however, due to the short exposure time, typically 1/20 second or less, the resulting total dose equivalent is small [approximately 0.2 mSv (20 mrem)]. On the other hand, if an unshielded radioactive source produces a radiation level in a work area of 0.2 mSv (20 mrem) per hour, an individual who works in that area for 40 hours per week would receive approximately 8 mSv (800 mrem) for each week worked. Thus, the amount of radiation received can be controlled by controlling the time of exposure.

7.2.2 *Distance*

If the distance from a point source of radiation is doubled, the exposure is quartered. (*i.e.*, a person standing 4 meters from an x-ray source will be exposed to only ¼ as much radiation as a person standing 2 meters from the source). This relationship describes the Inverse Square Law: The exposure rate from a point source of x or gamma radiation is inversely proportional to the square of the distance from the source. Thus, the amount of radiation received can also be controlled by controlling the distance from a source of radiation.

Radiation that scatters can also be a source of radiation exposure. As an x-ray beam passes through a patient, some of the x rays interact and change direction, or scatter. At one meter from the patient at a right angle to the beam, the scattered radiation intensity for x rays of the energy generally used in diagnosis is approximately 0.1 percent (0.001) of the intensity of the beam incident on the patient. This percentage may increase somewhat at angles greater than, or less than, 90 degrees. All personnel involved in x-ray procedures *should* stand as far as possible (at least two meters is desirable) from the x-ray tube and the patient and behind a shielded barrier or out of the room whenever possible. (See NCRP, 1981 for recommendations for the neonatal intensive care nursery.) Refer to Section 8.4.8 for further recommendations on use of fluoroscopic and cine equipment.

7.2.3 *Shielding*

Radiation interacts with any type of material and the amount of radiation is reduced on passage through materials. Thus, materials can be used to shield against radiation. However, some materials (*e.g.*, lead or concrete) make more efficient shields than others. Generally, in choosing shielding one must consider the type and energy of radiation involved and these are considered in constructing shielding heads for equipment and shielding walls for radiation rooms. They need to be considered also for such things as shielding aprons.

Lead aprons are efficient shields at typical diagnostic x-ray energies; however, they are not as effective for shielding the higher energy x-ray or gamma-ray emitters often used in nuclear medicine or radiation therapy and in some radioisotopes research laboratories. Beta particles convert some of their energy to x rays upon interaction with matter. The higher the atomic number of the absorber the greater is the percentage of energy converted to x rays. For this reason, when working with relatively large quantities of high-energy beta emitters such as phosphorus-32 (^{32}P), low atomic number shields (e.g., plexiglas) are often used. Stock solutions of ^{32}P in use or in storage are shielded with lead of sufficient thickness to absorb the x rays produced when the beta particles interact with their containment vials.

7.3 Survey Meters

Correct use of properly calibrated and maintained survey instruments is essential for detection and measurement of radiation in the workplace (NCRP, 1978b). Commercially available, sensitive, portable radiation detectors can provide rapid indication of the presence of radiation or radioactive materials or the adequacy of shielding. Individuals who are involved in the handling of radioactive materials *shall* be competent in the use of survey meters.

There are several types of portable survey meters in use in medical facilities, including the Geiger-Mueller (GM) counter, portable scintillation detectors and the ionization chamber.

GM counters normally overrespond to low energy x rays and, unless specifically calibrated for the energy of radiation being detected, these instruments should not be used for *quantitative* radiation exposure measurements. These instruments are, however, very sensitive and are useful for the qualitative detection of low levels of radiation. GM counter response should normally be recorded in counts per minute (cpm) rather than in dose rate or exposure rate, μGy (mR)/h. Calibrated ionization chamber instruments are much less energy dependent and give a more accurate indication of the exposure (ionization produced in air through interactions of radiation). Portable scintillation detectors are very sensitive for detecting low energy photons such as those emitted from iodine-125 (^{125}I). The response of radiation detection instruments *shall* be checked periodically using an appropriate radiation source. In measuring or detecting radiation, it is important to choose the proper instrument. The RSO *should* be consulted prior to the procurement of survey meters.

7.4 Personnel Monitoring Devices

Personnel monitors are small devices that can be worn by an individual for the purpose of estimating exposure to radiation. Examples of personnel monitors include film dosimeters, thermoluminescent dosimeters (TLD), pocket ionization chambers, and other small radiation detection devices. Personnel monitoring devices should be worn on the body (*i.e.*, collar, waist, etc.) as institutional policy dictates.

Most dosimeters measure a time-integrated dose, and the findings are reported as the effective dose equivalent for the period of use. This dose represents only an estimate of the effective dose equivalent to the body. If a dosimeter is worn at the collar, outside a lead apron, it represents only an estimate of dose to the head and neck. Personnel dosimeters generally will not record doses less than 0.1 to 0.2 mSv (10 to 20 mrem) and lower doses are often recorded as "M" for minimal (below detectable level). Dosimeters often contain filters which enable differentiation between penetrating and non-penetrating radiation. The site where the dosimeter is worn *should* be documented in the records. The RSO should be contacted for guidance on the appropriate personnel monitor to use.

7.5 Radioactive Materials Labels, Signs and Warning Lights

In addition to time, distance and shielding, contamination control measures also *should* apply to the use of radioactive materials. Radioactive materials *should* be restricted to authorized locations and *should not* be allowed in areas where its presence can be an unsuspected source of radiation exposure to personnel. For the most part, handling radioactive materials involves common sense and simple procedures, such as outlined in Section 8.3, which are not unlike the precautions applied to infectious disease. Individuals need to be warned of all areas containing radiation sources through appropriate signs, labels or other warning systems.

Warning signs *should* be used for radioactive material containers, radiation-producing devices, laboratories and other areas in which radioactive materials or radiation producing devices are used or stored. Signs may vary in shape or wording content; however, they all *should* contain the recognized magenta radiation symbol on a yellow background [black on yellow is also considered acceptable (ANSI, 1979)]. These signs indicate a potential hazard. Employees should be cognizant of such signs, and *should* follow explicitly any instructions listed on the signs. Instructions need to be in "plain" language and unambiguous. In certain areas where there is potential for high radiation levels, additional warning and protective devices such as lights, bells or interlocks may be installed to prevent inadvertent entrance. Specific guidelines for providing warning and access control are found in NCRP Report No. 88 (NCRP, 1987c).

Employees who have occasion to enter such areas *shall* be given appropriate training and safety instruction by a qualified individual. All such instructions or directions *shall* be followed by all employees.

7.6 Acquisition, Storage and Disposition of Radioactive Materials

In order to maintain control of incoming sources and provide for effective practices in their use, the following are recommended.

Purchase orders for radionuclides *should* be approved in advance by the RSO or designee. Only authorized users *shall* be allowed to order and receive radionuclides. Each authorized user *should* maintain an inventory of all radioactive materials in the user's possession. The type of radionuclides and the amounts specified on individual authorizations *shall not* be exceeded.

All incoming shipments of radioactive materials *shall* be delivered to a designated receiving area. It is important that the packages be checked for contamination, and exposure levels determined upon receipt. Night and weekend deliveries *shall* be taken to a designated receiving site, and security guards or other authorized personnel *should* accompany the carrier to this receiving site. Such security guards *should* be trained to determine if the condition of the shipment necessitates notification of the RSO for instructions.

All radioactive materials *should* be stored in a secured (locked) area. The need for shielding is determined by the type and energy of radiation emitted and the activity level of the radioactive material. Special handling and storage is determined by the type, energy and physical or chemical form of the radioactive material. Food *shall not* be stored in the same refrigerator or freezer used for storage of radioactive materials (or toxic chemicals or biological materials). Eating, drinking and smoking *shall* be prohibited in areas where radioactive materials are stored or used.

The RSO *should* inform all authorized users of the specific records they *should* keep and the procedures they *should* follow in preparing and disposing of their radioactive waste.

All authorized users, and potentially involved ancillary personnel, *should* be aware of the institution's emergency procedures (see appendix A) that are to be followed in the event of a spill or unexpected event.

The RSO *should* supervise all shipments of radioactive materials leaving the facility to ensure proper documentation and compliance with transportation regulations.

7.7 Radioactive Waste Management

An acceptable method for safely disposing of radioactive waste is required for anyone using radioactive materials. Radioactive waste to be shipped from the institution *shall* be appropriately packaged and managed so that it presents no hazard in transportation. Radioactive wastes stored in the institution require the same attention as given to use of radioactive materials. The RSO and the institution's policy manual *should* be consulted before making decisions on the disposal of radioactive materials. There are a number of options available to medical facilities for disposing of radioactive waste. These options are discussed below.

Storage for Decay

Most radioactive material used in medical applications is short-lived and will decay to background radiation levels within a short period of time (days to months). Therefore, space is generally made available for storage of radioactive waste on site. This is a very cost effective method for disposal of radioactive waste. Storage is also a solution to the handling of contaminated bedding, clothing and equipment. Usually, after a sufficient storage period (10 half-lives), and following surveys with appropriate instrumentation, these materials can be classified as non-radioactive and released from restriction after removal or destruction of all radioactive material warning signs or labels.

Shipment for Burial

If the radioactive material has a half-life that makes it impractical to store for decay, or if the facility has inadequate storage space, then shipment to a licensed commercial radioactive waste disposal site is required. Medical facilities often contract with a radioactive waste "broker" for periodic pick-up of their radioactive waste. The broker, in turn, stores the radioactive waste at an approved local facility until sufficient volumes are accumulated to make full truck-load shipments to a site licensed for the disposal of radioactive materials. The RSO *should* supervise the packaging of waste for shipment, and certify compliance of the shipment with the requirements of the disposal site and the regulatory agencies.

Release to the Environment

Certain radioactive materials may be released to the environment provided they are released in concentrations that are considered acceptable. In nuclear medicine procedures involving inert radioactive gases and aerosols, some release to the environment is unavoidable. Ideally, releases to the environment will be made from locations that are inaccessible to personnel, and that provide extensive dilution before the released material reaches any occupied areas. Planned releases *shall* be reviewed and approved by the RSO and *shall* be in accordance with federal, state and local regulations.

Incineration of radioactive wastes is acceptable but usually requires federal and/or state approval.

Release to the Sanitary Sewer

Certain water-soluble radioactive materials may be released to the sanitary sewer provided they do not exceed authorized release concentrations. A record of such releases *shall* be maintained. Patient excreta are generally considered acceptable for release to the sanitary sewer. Unless otherwise specified by the RSO, patient excreta may go directly into the sanitary sewer without consideration of radioactive content.

Special Consideration

In the hospital setting, care *should* be exercised to ensure that radioactive and infectious wastes are segregated and kept away from ordinary trash. If waste is both radioactive and infectious, it *shall* be sterilized before disposal as radioactive waste. Radioactive waste *should* be inspected for unwanted materials such as tissue or infectious material prior to storage for decay.

8. Guidelines for Specific Personnel

The previous sections provided general information that should be reviewed by all personnel involved with radiation. The following guidelines provide specific information for personnel who may be involved with radiation sources. Each subsection is independent and provides the necessary information for each designated group.

8.1 Administrators

8.1.1 *Responsibilities and Authority*

Management has the responsibility for ensuring the establishment of a radiation safety program including staffing level, staff and equipment and operating budget, written policies, procedures, and instructions that will foster radiation safety within the institution. In addition to the appointment of a Radiation Safety Committee (RSC) and a Radiation Safety Officer (RSO), management *shall* ensure that a formal annual review of the entire radiation safety program is performed. The review *should* include operating procedures (*i.e.*, the radiation safety manual or any other written policies dealing with radiation safety), past exposure records, results of investigations of any unusual radiation exposure incidents, inspections and recommendations of the RSO. See Report No. 59 (NCRP, 1978a).

8.1.2 *Implementation*

Administrators normally will choose to delegate to the RSO and the RSC responsibility for day-to-day supervision of the radiation protection program. Responsibility for safe practices with radiation sources lies with all personnel involved with the activity, *i.e.*, the principal investigator, the department head, the research technician, the technologist, etc., and management should assist in instilling

34

this view in all personnel. Deviations from safe practice must be reviewed and corrected immediately.

Radiation Safety Committee

The RSC is responsible for developing and maintaining an effective radiation safety program. The RSC meets periodically to review the radiation safety program and to make decisions on specific radiation safety questions or problems that are presented by the RSO or others. The committee *should* require the submission of a written radiation safety analyis of proposed programs and operations involving radiation and/or potential contamination. Proposals *should* be prepared by or be authorized by the radiation user and *should* detail standard operating procedures that will be used in both normal and emergency situations. Such proposals *should* be reviewed by the RSO prior to submission to the RSC. The RSO and the RSC *should* be available for and *should* respond to comments, questions or suggestions from all personnel. The RSC *should* review scheduled radiation safety briefings and educational and training sessions presented by the RSO to authorized users, workers, and ancillary personnel and approve the contents of the presentations.

All deviations from good safety practices *shall* be investigated immediately by the RSO and appropriate corrective measures *shall* be initiated. In instances where the RSO's authority to initiate corrective action is challenged, management *should* intervene. When worker health and safety are involved, the RSC *shall* take whatever steps are necessary to correct the situation, including immediate cessation of usage of radiation sources and revocation of the user's authorization.

Delegation of Authority

Management should delegate to the RSO the authority to inspect and enforce the radiation safety program. Management should support the RSO in any instance where it is necessary to assert authority. The RSO *shall* keep the RSC and management apprised of any problems, unusual events, or unjustified exposures. Management *should* ensure immediate compliance with all appropriate recommendations from the RSC.

Financing and Staffing

Sufficient funding, personnel and space to support the staffing and operation of the Radiation Safety Office and the purchase of supplies and equipment necessary to execute the Radiation Protection Program *should* be provided by the administration.

8.2 Animal Care Personnel

8.2.1 *Education*

Animal care personnel encounter the same need for protective measures in connection with radiation use on research animals as hospital employees do with patients (NCRP,1970c). Animals have an important role in certain research programs involving principally the use of radioactive materials, but on occasion involving exposure to x rays or portable beam sources. Animal care personnel *shall* be thoroughly trained in the appropriate protective measures required for the safe use of such sources of radiation.

8.2.2 *Signs*

Animal care personnel *shall* be aware of the significance of the various radiation protection warning signs and labels and follow any written precautionary measures that may be included on such signs.

8.2.3 *Waste*

Frequently animal care personnel are involved with handling of excreta, bedding or carcasses of animals that have been used in radionuclide research projects. Specific precautions *shall* be taken to control contamination from these types of waste and to ensure their proper disposal. In most instances, specific written instructions will be issued by the Radiation Safety Officer (RSO) on a form similar to that illustrated in Figure 8.1.

Waste from such animal experiments *shall* be clearly labeled with the radionuclide and the activity contained. Such waste must be handled as radioactive waste rather than being disposed of in the normal fashion. Animal care personnel *shall* place the waste in the designated and labeled containers, freezers or walk-in cold rooms and notify the RSO of its presence.

8.2.4 *Necropsy*

Necropsy of animals containing radionuclides can present both contamination and irradiation hazards. These should be discussed with the RSO when the experimental protocol is received (see Figure

ANIMAL-RADIONUCLIDE EXPERIMENT APPROVAL

INVESTIGATOR _____

LAB USED _____

TYPE OF ANIMAL USED _____

NUMBER PER EXPERIMENT _____

RADIONUCLIDE _____

RADIOCOMPOUND _____

DOSAGE PER ANIMAL _____

FREQUENCY OF EXPERIMENT _____

EXPECTED ROUTE OF EXCRETION _____

TIME UNTIL SACRIFICING _____

SAMPLE TAKEN _____

EXPECTED DATE OF COMPLETION _____

PROCEDURE FOR HANDLING EXCRETA AND BEDDING ____

_____

PROCEDURES FOR CARCASS AND WASTE DISPOSAL _____

_____

SPECIAL INSTRUCTIONS FOR ANIMAL CARETAKERS _____

_____

NECROPSY PRECAUTIONS _____

_____

COMMENTS _____

_____

DATE OF RSO APPROVAL _____

APPROVAL BY _____

Fig. 8.1 Example of animal-radionuclide experiment approval form.

8.1) and procedures for radiation protection, sample collection and waste disposal should be specified in advance.

8.2.5 *Records*

It is important that animal handlers pay particular attention to any records that are required for radionuclide animal experimentation projects. In particular, this applies to domestic farm animals that are not sacrificed after the experimental project is concluded. Proper steps need to be taken to ensure that such animals are not inappropriately used for human consumption.

8.2.6 *Irradiation Procedures*

The safety precautions dealing with irradiation of animals are basically the same as for irradiation procedures with human patients. It is important for the protection of individuals involved that close adherence to the recommendations of time, distance, shielding, personnel monitoring, restraining devices, etc., be utilized. Animals *should* be held using mechanical restraining devices. If such devices cannot be employed, the handlers *shall* wear a lead apron and gloves and take all reasonable precautions to avoid exposure to the direct radiation beam.

8.3 Clinical/Research Laboratory Personnel

8.3.1 *Introduction*

Radiation safety in the clinical or research laboratory cannot be separated from the general rules of laboratory safety; however, the risk of contamination and exposure from radioactive materials does require additional awareness of the possible hazards and unique monitoring requirements. Radioactive materials are used in virtually every phase of biomedical research. The radionuclides commonly used in medical research are shown in Table 8.1. The basic safety measures common to all research efforts will be discussed below.

TABLE 8.1—*Radionuclides commonly used in biomedical research*

Radionuclide	Half-life	Type of Radiation
^{3}H	12.3 y	Beta -
^{14}C	5730 y	Beta -
^{32}P	14 d	Beta -
^{35}S	87 d	Beta -
^{51}Cr	28 d	Gamma
^{59}Fe	45 d	Beta - ; Gamma
^{111}In	2.8 d	Gamma
^{125}I	60 d	X Ray; Gamma
^{131}I	8 d	Beta - ; Gamma

8.3.2 *Monitoring Requirements*

The use of radioactive tracers in medical research has increased significantly in the last few years. Some of these radionuclides have long physical half-lives (*e.g.*, tritium - 12.3 years; carbon-14 - 5730 years). Thus, contamination will not rapidly disappear and radionuclides retained in the body will continue to irradiate tissue for some time. However, residence time in the body is also influenced by the chemical form and metabolism of the particular compound retained in the body. Geiger-Mueller instruments with end window or pancake probes easily detect carbon-14. Special meters are available for detecting tritium contamination on surfaces; however, wipe testing remains the principal method of testing for tritium contamination. Monitoring of personnel for radiation exposure with a film badge or thermoluminescent dosimeter (TLD) is of no value for tritium and carbon-14, because these radionuclides emit only very low energy beta particles and present no external radiation hazard. Similarly, most survey meters cannot detect the low energy beta radiations from tritium or ^{14}C, although survey meters are effective for detecting higher energy beta particles.

Radionuclides, if internally deposited, are a concern and bioassay guidelines have been developed for monitoring for ingested or inhaled radionuclides (NCRP, 1976b; 1987b). Practices and procedures which could lead to the ingestion or inhalation of radioactive material *should* be avoided.

The use of MBq (mCi) or greater amounts of ^{125}I or other nuclides such as tritium for synthesis or labeling of compounds is potentially hazardous. These activities may lead to significant intake or external contamination if caution is not observed. Special procedures and protective devices can be used to minimize exposure and intake. Thyroid uptake counting can be used to determine intake of radioactive iodine and should be used as a bioassay method following iodination procedures.

8.3.3 *Education and Training*

Education and training in proper radiation safety practices in the laboratory are essential. Institutions engaged in large radionuclide research projects should have organized education programs for their staff. Such programs *should* be supervised by the Radiation Safety Officer (RSO) and *should* outline the basic radiation protection policies of the institution and the specific guidelines that are expected in all laboratories. This training *shall* be documented and *should* include a discussion of radiation risks and a description of the overall exposure profile for the institution. Such training *should* be supplemental to the specific training provided by the laboratory director or principal investigator. Written material *should* be provided which can re-emphasize the material covered in the training. The training program *should* provide for the diversity of radionuclide usage and should also provide for interim training of individuals hired between formal training programs. Requests for seminars should be addressed to the radiation protection staff. Additional information on radiation safety training is available in Report No. 71 (NCRP, 1983a).

8.3.4 *Area Designation*

The RSO should review each laboratory space where radiation sources are used and should decide what security precautions and warning labels are appropriate. Usually the entrance to laboratories using radioactive material should be posted with a sign bearing the radiation caution symbol and the words "CAUTION: RADIOACTIVE MATERIAL". Specific information on access control and appropriate signs can be found in NCRP Report No. 88 (NCRP, 1987c).

8.3.5 Precautions

The risk of equipment and surface contamination in a laboratory using radioactive materials demands extra caution by all individuals working within the laboratory to prevent release of radioactive materials from containment and avoid unnecessary exposure. Personnel *shall* adhere to all laboratory safety regulations. The following are general instructions for a radionuclide laboratory:

1. Wear protective knee-length lab coats and impermeable gloves whenever contamination is possible.
2. Use pipette filling devices for all procedures requiring the use of pipettes. *Do not pipette radioactive liquids by mouth!*
3. Use fume hoods and good contamination control principles when working with potentially volatile radionuclides.
4. Survey hands, shoes, and clothing for contamination using an appropriate instrument designated by the RSO before leaving the radioisotope laboratory.
5. Do not drink, eat, apply cosmetics or smoke in areas designated as restricted for the use of radioactive materials.
6. Survey the working area for contamination and proper storage of radioactive materials during and after the use of radioactive materials.
7. Cover work surfaces with plastic-backed absorbent material to limit and collect spillage in case of an accident. Keep work areas clean, orderly and free of extraneous material.
8. Label radioactive material containers as directed and store such containers in the designated storage area.
9. Label and isolate radioactive waste and equipment. Once equipment has been used for radioactive substances, it *should not* be used for other work nor *should* it be removed from the area until it is cleaned and surveyed to demonstrate the absence of radioactivity.
10. Immediately report all accidents involving radioactive materials to the supervisor and the RSO. Be aware of decontamination procedures and take whatever steps are necessary to prevent spread of contamination.
11. Maintain records of all radioactive materials delivered to the laboratory and the use and disposal of these materials.
12. Wear any personnel monitors that are required by the RSO.

8.3.6 *Waste Disposal and Storage*

Each laboratory *shall* be equipped with special waste containers appropriately labelled for solid, liquid, animal or biological radioactive waste. Separate containers *should* also be designated for separation of short-and long-lived radioactive wastes. All radioactive waste *should* be transferred to the RSO for disposal in accordance with appropriate regulations. It may be permissible to dispose of small quantities of liquid radionuclides in the sanitary sewer system provided the RSO has approved the procedure and appropriate records are maintained. Radioactive waste *shall not* be discarded in the normal trash unless surveyed and approved for disposal by that method by the RSO.

Containers used for shipping or storage *shall* be labeled with the radiation caution symbol and the words "CAUTION: RADIOACTIVE MATERIAL". The label *shall* also state the quantities and kinds of radioactive materials in the containers and the date of measurement. Empty containers *should not* be disposed of as nonradioactive waste unless the containers are free of contamination and the radioactive material labels are removed or disfigured.

8.3.7 *Animal Research*

The use of animals in radionuclide research will not be a source of contamination or personnel exposure if the researchers have taken adequate precautions and anticipated potential problems before initiating their work. Special consideration is required for the possibility of contamination from urine, blood, saliva and fecal matter. Animal carcasses *should* be sealed in a plastic bag that is properly labelled and stored in a designated cold area for disposal as directed by the RSO. Laboratory personnel using animals in research should be familiar with the contents of the Section for animal handlers.

8.3.8 *Emergency Procedures*

Even in the most controlled setting, radiation-emitting materials will inadvertently be released and an area will be contaminated. It is essential for the user to be familiar with the emergency and/or decontamination procedures to be followed while awaiting the arrival of the RSO. Each laboratory *should* have emergency and decontamination procedures posted in a conspicuous position. Sample emergency and decontamination procedures are contained in Appendix A.

It is essential that every radionuclide user be capable of evaluating the risk of contamination and exposure from accidental spills in the laboratory. Minor spills can be decontaminated by laboratory personnel followed by a report to the RSO. The area *should* be surveyed to confirm that decontamination efforts were successful. When in doubt, notify the RSO. In the event of a major spill or personal injury, the RSO *shall* be notified immediately. In either case, immediately notify other users in the area regarding the incident and prevent the spread of the contamination by covering the spill with absorbent material. Subsequent procedures will depend on the magnitude of the spill and are outlined in Appendix A. Information on actions to be taken in the event of personnel contamination or injury can be found in Report No. 65 (NCRP,1980b).

Personnel decontamination in the event of a minor incident *should* consist of removing and storing any contaminated clothing. Minor skin contamination *should* be removed by flushing the area and washing thoroughly with a mild soap and cool water. Care should be taken not to abrade or inflame skin surfaces during this procedure. Major decontamination procedures *should* be supervised by the RSO and appropriate medical personnel. In the event of trauma associated with an accident, it should be kept in mind that vital first aid or life saving measures take priority over contamination control. See NCRP Report No. 65 (NCRP, 1980b) for more detailed information on the subject.

8.4 Diagnostic X-ray Technologists

8.4.1 *Introduction*

Each year, radiologic technologists perform over 180 million radiographic procedures on patients in the United States (NCRP, 1989a). In order to protect the patients and themselves, the technologists should employ the practices and principles discussed in this section.

Many technologists are involved in diagnostic x-ray procedures including radiography with mobile equipment, fluoroscopy, tomography, mammography, computed tomography, angiography, cardiac catheterization, and interventional radiography. These areas will have protection considerations unique to the equipment and procedures used at each medical facility. The technologist *should* be thoroughly familiar with the equipment and its operating procedures. Training is a necessity. The technologist *should* periodically consult the Radiation Safety Officer (RSO), radiation physicist, and the radiologist or chief technologist to ensure that proper radiation safety practices are being followed.

8.4.2 *Education*

The responsibility for the proper use of radiation-producing equipment ultimately lies with the user (in this case, the technologist). Every medical facility *shall* have an orientation program on radiation safety for newly employed technologists and a continuing education program to review the rules and regulations.

Certain government agencies have issued recommendations and regulations pertaining to the proper use of ionizing radiation. In the United States, the federal government, acting through the Center for Devices and Radiological Health (CDRH), has mandated compliance standards for the manufacturing of x-ray systems (FDA, 1986), but the federal government does not regulate the user. Most states have developed regulations pertaining to safe use of radiation-producing equipment. Technologists *should* have access to and be familiar with these regulations.

As a result of the educational program, the technologists *shall* be aware of the approximate amount of radiation received by their patients during each radiographic procedure used in their facility. They *should* also be informed of the approximate amount of radiation they are likely to receive for a normal workload within their assigned working areas.

8.4.3 *Equipment Operational Procedures*

As a general rule, any technique or procedure that reduces the number of retakes and patient exposure reduces the exposure to radiologic personnel. Technologists are responsible for knowing and using proper techniques, positioning, image receptor (screen-film combination, image intensifier, image mode, etc.), appropriate shielding, and collimation, with each patient. It is imperative that the technologist understand the effects of all technique factors on overall image quality. It is also important for the technologist to understand the differences among the types of image receptors used in radiology and the effects of these different image receptors on patient and personnel exposure. For example, a rare earth screen-film system can reduce patient exposure significantly compared with conventional screen-film combinations.

The considerations related to proper use of the equipment that minimize operator and patient exposure are discussed in the following sections.

Proper Collimation

As the size of the beam is decreased, the number of scattered photons is also decreased; therefore, proper collimation minimizes both the volume of tissue irradiated in the patient by direct x rays and the number of photons available for scatter to the operator. There is a significant reduction in scattered radiation for small fields as compared to large fields.

The maximum allowable dimensions of the x-ray beam *shall* never exceed the size of the image receptor for both radiographic and fluoroscopic procedures. In order to maximize image contrast and minimize exposure to the patient and staff, the x-ray field should be collimated to the body part of interest, even if that field is significantly smaller than the image receptor.

Proper Filtration

Filtration removes the low-energy photons from the primary x-ray beam thus producing a beam with a higher effective energy and increased average penetration. The low-energy photons which are absorbed superficially within the patient are a primary source of unnecessary patient exposure and tend to reduce image contrast. Proper filtration removes most of these photons before they have the opportunity to interact with the patient.

Technique Factors

The ultimate clarity of a radiographic image is a complex product of many factors including subject contrast, radiographic contrast, fog and scatter, sharpness, radiographic mottle, and resolution. Each of these factors has one to several variables which directly or indirectly alters its effect on overall radiographic image clarity. Many of the factors which ultimately determine the quality of the image are under the direct control of the radiologic technologist. Parameters such as kVp, milliampere second (mAs), grid selection and image receptor are routinely determined by the technologist. It is essential that technologists be aware of the effects of variations in these parameters on image quality and patient exposure and ensure that they have selected the correct parameters for each study. Techniques *shall* be provided for each x-ray system clearly stating the kVp, mAs, grid and image receptor type for each examination performed in that room, and for various body part thicknesses.

High-Speed Image Receptor System

As the speed of the receptor system is increased, the number of photons required to produce an image decreases. The technologist should be aware of the effects of different screen-film combinations on patient exposure and image quality. In general, the fastest system which will not compromise image quality is optimal for patient and operator safety.

Orientation of the Beam

The x-ray beam *should* be directed so as to achieve minimal exposure to everyone and optimal image quality on the radiograph. In a radiographic room, the technologist *should* be aware of the shielding design of the room and *should* use the equipment accordingly. Careful orientation of the beam, especially for oblique and lateral films so that the direct beam cannot strike people, is essential in protecting operating personnel, especially in radiography with mobile equipment. Beam centering devices *should* be used when available and the beam *should* be carefully collimated to the area of clinical interest.

Leakage Radiation

The maximum permissible leakage radiation through a diagnostic housing is limited by federal compliance standards to a safe level for normal operating distances. During certain procedures, such as radiography with mobile equipment and oblique cinefluoroscopy, the technologist *should* stand as far as practicable from the tube housing.

8.4.4 *Holding Patients*

Individual medical personnel *should not* have the responsibility of routinely holding patients during diagnostic radiology procedures. In particular, this *should not* be a practice routinely demanded of individuals who are designated as radiation workers (*e.g.*, the x-ray technologist). Patients *should* be held only after it is determined that available restraining devices are inadequate. Individuals holding patients for x-ray procedures *should* be provided with lead aprons and lead gloves and *should* be positioned so that no part of their body is exposed to the direct radiation beam. To assist in minimizing exposure, it is important for the radiologic technologist to collimate carefully to the area of clinical interest. Pregnant women or persons under the age of 18 years *should not* be permitted to hold patients. Actual guidelines or policies for the selection of the individuals responsible for holding patients at a particular institution *should* be reviewed and approved by the radiation safety committee within that institution.

8.4.5 *Shielded Booths*

The wall shielding in the radiographic suite *shall* be designed to protect the operating personnel, other employees and the general public. Most radiographic suites will have a shielded booth, and the operator *should* remain in this booth during the procedure.

Each time an x-ray beam scatters, the intensity of the beam at 1 meter from the scattering object is decreased by approximately 1000 times. In order to minimize operator exposure, the room *should* be designed so that no x rays can enter the shielded booth unless they have been scattered at least twice. This means that even if the distance factor is ignored, the intensity of the x rays in the shielded booth is about one millionth of the primary beam. Further, the exposure switch *should* be located to prevent energizing the x-ray tube when the technologist is outside of the control booth.

8.4.6 *Mechanical and Electrical Safety*

Although mechanical and electrical safety are the primary responsibility of installation and service personnel, technologists *should* be cognizant of the possibility of a mechanical or electrical hazard and report it immediately to the appropriate individual.

8.4.7 *Personnel Monitoring*

Film dosimeters, thermoluminescent dosimeters (TLDs) or pocket dosimeters are recommended for use by operators of medical x-ray equipment. The dosimeter should always be worn at the same general location on the body. The exact position of the dosimeter should be determined by the RSC for each situation.

In Report No. 91 (NCRP, 1987a) the NCRP recommends the use of effective dose equivalent for radiation protection purposes.

The positioning of the dosimeter varies from institution to institution, with some placing the dosimeter under protective garments, and others placing the dosimeter outside of protective garments. In other instances two dosimeters are utilized, one under and one outside of protective garments. The Radiation Safety Officer (RSO) or other qualified expert should consider the appropriateness of these alternatives for specific situations.

There are two situations of concern; one, when no lead apron is necessary and the other when the leaded apron should be worn. In the first situation, the dosimeter *should* be worn normally on the trunk of the body at waist level or above so that the part of the body likely to receive the greatest proportion of its dose equivalent will be monitored. When the apron is worn, a decision must be made as to whether to wear one or more than one dosimeter. If only one is worn and it is worn under the apron it can represent the dose to most internal organs but it may underestimate the dose to the head and neck (including the thyroid gland). If only one is worn, and it is worn at the collar, it may represent the dose to the organs contained in the head and neck but it may overestimate the dose to the organs in the trunk of the body. In either case, if only one dosimeter is worn, the RSO needs to be fully aware of the factor by which the particular placement underestimates or overestimates the effective dose equivalent so that an accurate estimate can be obtained. If two dosimeters are worn, the RSO needs to adapt a method for combining the two doses to yield an estimate of the effective dose equivalent.

Consideration *should* be given to the monitoring of pregnant personnel to assure that the dose equivalent to the fetus does not exceed 0.5 mSv (50 mrem) in a month (NCRP, 1987a). It is recommended that the dosimeter be placed at waist level and under any protective apron for this purpose.

8.4.8 *Fluoroscopy, Special Procedures and Cardiac Imaging*

The radiation level (air kerma) from a fluoroscopic unit may be as high as 87 mGy (10 R)/min measured at the patient skin entrance. There is, consequently, a high potential for radiation to be scattered from the patient, and certain precautions must be taken during fluoroscopy. This is especially true where more than one fluoroscopy tube may be in operation at any given time or when the procedures may be lengthy (*e.g.*, special procedures and cardiac imaging). A leaded apron *shall* always be worn; the screen-film slot cover, if present, and Bucky slot covers on under-table x-ray tube fluoroscopy units *should* be in position to protect the operator. Distance to the source *should* be maximized, leaded gloves *should* be used, and the lead drape attached to the image intensifier housing *should* be positioned to minimize radiation exposure. Each room *should* be specifically surveyed by the RSO who can prescribe special shields and procedures as indicated.

Radiologists and physicists need to be aware that the increasing use of dynamic CT scanning creates the potential for significant exposure of personnel due to radiation scattered from the patient. Steps to minimize such exposure should be considered (Kaczmarek *et al.*, 1986).

8.4.9 *Special Requirements for Mobile Equipment*

The cord to the exposure switch for mobile equipment *should* be of sufficient length to permit the operator to be at least 2m (6 ft) from the patient during an exposure. If the console is large enough to permit the operator to be adequately shielded by the console, then the cord to the exposure switch *should* be short to encourage the operator to remain behind the console. All operators of mobile equipment *should* wear lead aprons. The operator of mobile equipment *should* be cognizant of orientation of the beam with respect to any other patients, employees or other individuals in the room. Care *should* be taken to maximize source-to-skin distance in order to reduce exposure to the patient.

Reasonable protection for assisting personnel involved in radiographic procedures with mobile equipment will be provided if they:

- Do not permit themselves to be exposed to the direct beam.
- Remain at least 2m (6 ft) from the patient, the x-ray tube and the useful beam.
- Wear a leaded apron and gloves when holding a patient or when it is necessary to remain closer than 2m (6 ft) from the beam.

(See NCRP, 1981 for requirements in the neonatal intensive care nursery).

- Hold patients only when necessary and then infrequently.
- *Should not* hold patients if they, the assisting personnel, are pregnant.
- Follow any radiation safety instructions given.

8.4.10 *Dental*

Radiation protection for the operator of dental x-ray units includes consideration of the proper position of the operator during the exposure and restrictions on holding the film during the exposure (NCRP, 1970a). The exposure switch *should* be located in a position that will require the operator to stand behind a suitable barrier during each exposure. If it is necessary for the operator to remain in the room with the patient during the exposure, then the operator should stand as far as practical from the patient and outside the path of the useful beam [*i.e.*, avoid the area between the x-ray tube and the patient or on the opposite side of the patient where the x-ray beam exits from the patient]. Operators *should not* hold the film or the x-ray tube in position during the exam. The operator *should* ensure that the proper beam-limiting /beam-indicating device is on the radiographic head and that the proper filtration is part of the radiographic unit.

8.5 Escort Personnel

8.5.1 *Introduction*

Working in a hospital setting does entail some occupational risks, including exposure to radiation. However, the need for escort personnel to be exposed to radiation is minimal, and hospital regulations *should* ensure that any possible exposure would be minimal. The Radiation Safety Officer (RSO) *should* be available to evaluate any situation that may be of concern to escort personnel.

8.5.2 *Radiology*

In some hospitals, escort service workers may be assigned permanently to the diagnostic radiology department. They may be asked to wear a dosimeter to document any possible exposure that might result from being near sources of radiation, although no measurable exposure would be expected from working in this department. Patients exposed to x-ray beams are not made radioactive by the procedures, and thus, do not constitute sources of exposure.

8.5.3 *Nuclear Medicine*

Patients often need transportation from the nuclear medicine department to their rooms. They have had a small amount of radioactive material administered as part of a diagnostic study. Most of the radiation is absorbed by the patient's body and patients can be transported without undue concern for the escort's exposure. If an escort is permanently assigned to the nuclear medicine department, the wearing of a dosimeter, appropriately located, may be required; this would be determined by the RSO as a result of review of the department's procedures.

8.5.4 *Radiation Therapy*

Many hospitals use radiation as a treatment for cancer and various benign conditions. A patient being transported to or from the radiation therapy department is most likely receiving external radiation therapy, *i.e.*, treatment from a source of radiation outside the body. This radiation is produced by high-energy radiation machines, and some of it passes through the patient's body. The patient does not become radioactive from such treatments. In this respect, the treatments are similar to chest x rays or other radiographic procedures.

Other cancer patients are treated by placing radioactive materials within the body. These patients contain a source of radiation, and measurements are taken to determine the radiation exposure levels around the patient. Most of these patients remain in their room the entire time they contain the radioactive source, but, occasionally, they require transportation. The radiation therapy physician or the RSO should assist with the transport of these patients, unless exposure levels around the patient are determined to be insignificant. Escort personnel may be asked to assist with the transport. These patients can be identified by the "CAUTION - RADIOACTIVE MATERIAL" label on or just inside the patient's chart and on their room door. The patient may also wear a wristband containing the radiation symbol and words of caution. The universal sign indicating the presence of radiation is a magenta symbol on a yellow background (black on yellow is also considered acceptable). The medical person assisting the patient should inform the escort as to location of the radioactive sources (*e.g.*, head, abdomen). The escort *should* stay as far from the sources as possible, thereby minimizing his/her exposure while transporting the patient, unless advised otherwise by the medical person or the RSO. If possible, designated patient elevators *should* be used; the general public *should* be excluded if public elevators are used. The least crowded corridors *should* be selected for passage. The RSO *should* advise more specific routing if necessary.

If requests for transport of these patients are infrequent (once a month or less), the exposure received would be minimal. If frequent contact with these patients is required of specific escort staff, the RSO *should* consider having them wear dosimeters to record their exposure over a period of time. They *should* also receive general instructions about radiation and about protective measures to keep exposures as low as practicable.

The RSO *should* be available to discuss any concerns or questions escort personnel may have about exposure or about the presence of radiation in their work environment.

8.6 Housekeeping (Janitorial) Personnel

8.6.1 *Introduction*

Housekeeping personnel, although rarely considered radiation workers, do have to be aware of the possible presence of radiation sources. They may be requested to service an area where "RADIO-ACTIVE MATERIAL" or "RADIATION AREA" signs are displayed. Waste receptacles in laboratories are often marked for radioactive materials. Hospital radiation safety regulations should be specific enough to ensure that housekeeping personnel are not exposed to radiation without their full knowledge and proper training.

8.6.2 *Education*

Housekeeping personnel, especially those servicing the nuclear medicine, radiology, and radiation therapy departments, and the radiopharmacy and radionuclide research laboratories, *should* receive some general instructions about radiation, the means to detect its presence, the meaning of radiation signs, symbols and labels and protective measures for keeping personal exposure low. Any concerns for improperly shielded materials or waste mix-up in the laboratories should be reported immediately to the supervisor and subsequently to the Radiation Safety Officer (RSO). Housekeeping staff *should not* be expected to assess the safety or hazards of a questionable situation involving radiation. The RSO *should* be available to these workers whenever concerns are raised.

8.6.3 *Laboratories*

Many laboratories use radionuclides daily, but these materials are kept in shielded areas so that radiation exposure in the room is within safe limits. Radioactive waste, usually paper, glass and plastic materials contaminated through usage, *shall* be kept separate from regular waste. It *shall* be marked clearly as "RADIOACTIVE MATERIAL", with a magenta (or black) symbol on yellow background and should be labeled, "HOUSEKEEPING PERSONNEL DO NOT REMOVE". The waste *should* be handled by laboratory personnel and stored separately from regular hospital waste. Empty containers that bear radioactive material labels *should not* be discarded until the labels have been defaced or removed.

If a member of the housekeeping staff enters a laboratory displaying a "RADIOACTIVE MATERIALS" sign and encounters any liquid or solid spillage, the door *should* be locked and the incident reported immediately to the supervisor. The RSO or the director of that laboratory will be able to determine if the spill involved radioactive materials by measuring it with a survey meter or other analytic means.

Although it is not a routine practice, housekeeping personnel may be asked to assist with cleaning an area contaminated with radionuclides. This *should* be done only under the direct supervision of the RSO. Disposable gloves *should* be worn; hands and shoes *should* be monitored when finished. Wash water may be discharged through the sewage system, at the discretion of the RSO.

8.6.4 *Controlled Areas*

There are radiation-producing machines in the radiology and radiation therapy departments, but they emit radiation only when the machines are turned on. Safeguards *should* be in place to ensure that housekeeping personnel and other non-radiation workers do not enter these rooms when the machines are activated. For cobalt teletherapy facilties, the RSO *should* provide specific radiation safety instructions to housekeeping personnel.

Several areas of the hospital may be designated "RADIATION AREA" and *shall* be so identified. If entry to the area is restricted because of increased levels of radiation, cleaning procedures *should* be assessed by the RSO, and any special instructions *should* be discussed with the housekeeping supervisor.

8.6.5 *Patient Care Rooms*

Housekeeping personnel may also be restricted from cleaning the rooms of some patients being treated with therapeutic levels of radionuclides. This restriction ensures that no source of radioactive material is inadvertently swept up and disposed of improperly. A radiation sign *should* be posted at the door to identify these rooms. If there is any question as to whether the room is restricted, the nurse-in-charge *should* be asked for clarification.

Housekeeping personnel may be asked, at the discretion of the RSO, to assist with cleaning the room of a discharged patient who has been treated with liquid radioiodine. There may be traces of the radioiodine on the bed linens, sink, telephone, around the toilet and any other areas touched by the patient. The room *should* be surveyed by the RSO prior to clean-up to identify specific areas of contamination. All housekeeping assistance *shall* be supervised by the RSO. Disposable gloves and shoe covers *shall* be worn while working, and hands *should* be washed upon completion. The room *should* be surveyed again by the RSO to ensure that all traces of radioiodine have been removed. The survey *should* also include the hands and shoes of the housekeeper when the job is finished. Disposable waste and linens that are contaminated with radioactivity *should* be bagged and taken to the radioactive waste storage area by the RSO.

8.7 Maintenance and Engineering Personnel and In-House Fire Crews

8.7.1 *Introduction*

There may be many areas of a hospital that contain radiation-producing devices or radioactive materials. It is important for maintenance and engineering personnel and members of in-house fire crews to be aware of such areas and the means of avoiding radiation hazards in them. It is also important to make certain that in carrying out their functions, they do not make any changes that would interfere with any radiation protection barriers or devices.

The maintenance department must be aware of the precautions they need to take when entering or working in radiation usage areas of the hospital or research laboratories, and all maintenance work in radionuclide laboratories should be approved by the RSO before initiation. Many hospitals maintain an in-house fire crew that responds to all emergencies until the local fire companies can arrive. These groups *should* maintain emergency information pertinent to all areas in which radionuclides are stored or used with the periodic assistance of the Radiation Safety Officer. Yearly instruction from the Radiation Safety Officer (RSO) *should* be requested.

Although maintenance is mentioned throughout this section, the recommendations apply equally as well to engineering personnel and in-house fire crews. The hospital engineering staff, particularly clinical engineers may find additional radiation safety information of specific interest to them in the *CRC Handbook of Hospital Safety* (Ziemer and Orvis, 1981).

8.7.2 *Plumbing*

Maintenance personnel *should* be aware that small quantities of radioactive materials may be disposed of by way of the sanitary sewer. Therefore, sinks and drains from designated radionuclide laboratories *should* be considered contaminated until appropriate surveying or monitoring can be performed. It is the responsibility of the RSO to provide, upon request of maintenance personnel, a survey and any necessary instructions before removing, repairing or replacing plumbing in these laboratories. In many institutions, specific warning tags containing written instructions to maintenance personnel are fastened to the sink traps. However, the fact that the room or laboratory is designated or marked as a radioisotope laboratory *should* provide sufficient warning to maintenance personnel so that they will request specific instructions from the Radiation Safety Office before proceeding.

8.7.3 *Ventilation*

Maintenance personnel *should* also be aware that radioactive materials are occasionally released to the air. Therefore, it is possible for fume hoods, exhaust ducts, filters and blower assemblies to become contaminated with radionuclides. They should contain appropriate warning labels and the RSO should be contacted to provide monitoring and specific instructions before work begins in these areas. Any modification to air handling systems should be reviewed by the RSO before work is begun to ensure that air flow patterns will not adversely affect the radiation safety program.

8.7.4 *Modifications of Shielded Rooms*

Frequently maintenance personnel are involved in projects that may alter shielding in walls or cabinets of x-ray rooms, therapy rooms or radionuclide laboratories. Such shielding is installed to provide protection for employees and the public from radiation. Any voids or openings in the shielding can permit excessive radiation to escape. Also, substitution of shielding that is less than that which was specified can lead to unacceptable radiation levels. Maintenance personnel *should* notify the Radiation Safety Office before making any alterations to shielding or shielded areas to ensure that the alterations will be in compliance with safety specifications. It is also important that the Radiation Safety Office be aware of such modifications so that appropriate surveying can be done either during construction or after completion.

Maintenance personnel *should* be aware that shielded enclosures, which have been assembled in radionuclide laboratories (*e.g.*, lead caves, lead pigs or other shielded areas), may contain radioactive materials. Therefore, before opening, disassembly or removal, the Radiation Safety Office *should* be contacted for specific instructions or surveys.

8.7.5 *Mechanical/Electrical*

Rooms containing x-ray equipment as well as those in which radionuclides are used often have specific mechanical and/or electrical requirements. Maintenance personnel *should* request specific instructions from the Radiation Safety Office before making any changes or replacements to either mechanical or electrical systems in these areas.

8.7.6 *Heating and Air Conditioning*

Certain pieces of radiation detection or counting equipment will operate properly only within a narrow temperature range. It is very important that these specified temperatures be rigidly maintained for certain counting or imaging laboratories.

8.7.7 *Appliances*

Appliances, such as refrigerators and centrifuges, in radionuclide laboratories may contain contamination as a result of their use with radionuclides. Prior to being removed from the laboratories or being serviced, appliances *should* be surveyed and certified as uncontaminated by the Radiation Safety Office.

8.7.8 *Labels, Signs and Warning Lights*

Radiation warning signs and labels should be conspicuous. Such signs and labels are used to warn of areas containing radioactive materials or other radiation sources. Maintenance personnel should be familiar with these signs and obey any instructions given with them. In the event that specific written instructions are not present, the Radiation Safety Office *should* be contacted before work is done in an area so designated.

8.7.9 *After Hours*

Typically, some maintenance personnel are on duty at all hours, whereas radiation safety personnel are usually present only during normal working hours. During such off-duty hours, maintenance personnel *should* be alert to any unusual situations in areas that are designated as radionuclide laboratories. Any occurrence that could involve radionuclides or other radiation sources *should* be reported to the Radiation Safety Office immediately. Maintenance personnel *should* maintain a listing of all areas in which radioactive materials are used or stored and a listing of the individuals to contact in an emergency. Such listings *should* contain abbreviated safety instructions for maintenance personnel who would have to enter these areas during emergencies when supervisors or someone from the Radiation Safety Office may not be immediately available.

8.8 Nuclear Medicine Technologists

8.8.1. *Introduction*

The nuclear medicine technologist prepares and often administers prescribed amounts of radiopharmaceuticals for both diagnostic and therapeutic purposes (NCRP, 1982). To do this safely, the nuclear medicine technologist follows the basic principles of radiation protection that are presented in Section 7. Additional unique considerations for the nuclear medicine technologist are reviewed in this section.

8.8.2 *Education*

Orientation of a new nuclear medicine technologist *shall* be conducted by persons who are well versed in nuclear medicine and radiation safety. This training *should* involve a review of the specific techniques that are used for radiation safety within the department as well as proper operation of imaging equipment. This material is often supplemented by periodic lectures by the radiation safety staff. Regulations on radiation protection from the appropriate state and/or federal agency *shall* be made available to all nuclear medicine technologists.

Technologists *shall* be aware of the approximate radiation doses to patients from the procedures and be capable of responding to the patient's concerns. Technologists *should* also be aware of the exposures they are receiving and the range of exposures that are appropriate for the types and numbers of procedures performed. Continuing education lectures on radiation safety *should* be available as a part of an in-service education program or at local and regional professional and/or scientific meetings.

8.8.3 *Handling and Administration of Radioactive Materials*

Most activities involving the preparation of radioactive pharmaceuticals occur in the "hot lab". The "hot lab" may consist of several rooms in a radiopharmacy or a special section of one room in small hospitals. Radioactive materials delivered to the "hot lab" should be inspected for damage, surveyed for contamination and exposure levels, opened properly and calibrated. Records of receipt, disbursement, and disposal of all radionuclides are required by government and accrediting agencies (See Section 7.6).

The properly equipped "hot lab" contains a number of tools and devices which *should* be used to minimize personnel exposure. These include L-shaped ("body") shields, syringe and vial shields, remote handling tools, plastic gloves and plastic-backed absorbent paper, and fume hoods or other suitably ventilated enclosures.

A potential problem associated with the use of radioactive materials is internal contamination, that is, inhalation or ingestion of radionuclides. For this reason, and to avoid contamination of work surfaces, extra care must be exercised to avoid release of radioactive materials to room air. Procedures involving potentially volatile radionuclides *should* be performed in a properly designed and operating fume hood.

Material dispensed from the "hot lab" to other areas of the institution *shall* be in suitable containers that restrict radiation exposure to administratively defined acceptable levels. These containers *shall* carry the required radioactivity labels. The container for radioactive material may be a lead container or merely a large fiber box, whose function is to ensure that there is sufficient distance from the radioactive material in an inner container. Therefore, it is imperative that persons handling these containers not alter them.

A special area *should* be designated for the injection of patients to limit the possibility of contamination to a small area and to minimize any effects this procedure might have on images being accumulated on other patients. When it is necessary to administer doses for diagnostic studies in other areas, such as intensive care units and cardiac stress laboratories, the radioactive materials shall be transported to and from these areas in shielded containers. Materials that become contaminated outside the department *shall* be returned to the nuclear medicine department for proper disposal. Special precautions *should* be taken while transporting contaminated material back to the nuclear medicine department.

Periodic surveys for radiation or contamination *should* be performed with an appropriate survey meter and paper wipes. The frequency of surveys in a given area is determined by the quantity, type and frequency of radionuclide procedures. Surveyed areas *should* be indicated on a drawing of the area that also contains space for the recording of results, the name of the surveyor and the instruments used in performing the survey. Wipe tests *should* also be performed because low levels of surface contamination may not be detected with a survey meter.

8.8.4 *Special Considerations Relating to Therapeutic Administration*

Therapeutic uses of radiopharmaceuticals usually involve large amounts [GBq (mCi)] of radionuclides and may involve potential hazards that warrant special consideration. Restrictions placed on the patient will depend on the amount and nature of the radionuclides involved (see Appendix B). The nuclear medicine technologist *should* be thoroughly familiar with the unique or increased precautions associated with therapy procedures (NCRP, 1970b; NCRP, 1972).

8.8.5 External Exposure Rates

Within the nuclear medicine department, 30 to 80 percent of the exposure to technologists comes from the imaging process, provided proper radiation safety procedures and shielding devices are used in preparing and administering the radiopharmaceuticals (Barrall and Smith, 1976). Table 8.2 presents dose equivalent rates near patients to whom radiopharmaceuticals have been administered. The data from the table can be helpful in evaluating the potential for exposure in dealing with patients who are undergoing diagnostic studies. A survey of the exposures of nuclear medicine technologists indicates that only a small percentage receive exposures of more than 10 percent of the permissible yearly dose limit (Lis *et al.*, 1981).

8.8.6 Holding Patients

Patients are usually not held during nuclear medicine procedures. In most cases, except for small children, tape or various devices such as padded weights can be used to immobilize the patient (NCRP, 1981).

8.8.7 Portable Shielding

Leaded Lucite™ or lead shields on wheels can be moved from one location to another and provide significant reduction of radiation exposure from radionuclides such as technetium-99m (^{99m}Tc). Lead aprons routinely used in radiology are not designed for protection against the higher energy gamma rays emitted by radionuclides used in nuclear medicine. It is normally more effective to shield the radiopharmaceuticals themselves. In particular, this is true of radiopharmaceuticals in vials or syringes (Lis *et al.*, 1981; Syed *et al.*, 1982). In addition to shielding, reducing exposure time and maintaining adequate distance are important safety practices in nuclear medicine.

TABLE 8.2—Data on dose equivalent rates from adult patients. (Syed et. al., 1982; Castronovo et. al., 1982; Jankowski, 1984).

Study	Radiopharmaceutical	Administered amount		Time after administration	Distance (cm)	Dose equivalent rate	
		MBq	(mCi)			μSv/h	(mrem/h)
Bone	99mTc MDP	740	(20)	0	100	9	(0.9)
				1 h	100	6.3	(0.63)
				2 h	100	4.7	(0.47)
				3 h	100	3.5	(0.35)
Liver	99mTc S colloid	150	(4)	0	100	2	(0.2)
Blood pool	99mTc RBC	740	(20)	0	100	14	(1.4)
Tumor	67Ga citrate	110	(3)	0	100	3.5	(0.35)
CSF	111In DTPA	19	(0.5)	0	100	0.8	(0.08)
Heart	201Tl chloride	740	(20)	0	a	20	(2)
Heart	99mTc HSA	190	(5)	0	a	15	(1.5)
Bone	99mTc MDP	740	(20)	0	a	25	(2.5)
Heart	99mTc RBC	1000	(27)	20 min	100	18	(1.8)

aSide of stretcher

8.8.8   *Emergency Procedures*

In the event of an accidental release or spill of any radioactive material, the Radiation Safety Office and the nuclear medicine physician *shall* be notified immediately (see Appendix A). However, a nuclear medicine technologist *should* be familiar with the principles of decontamination and *should* be capable of handling any emergency involving radionuclides used in nuclear medicine. Emergency instructions and names and numbers of individuals on the emergency call list *should* be posted in convenient locations.

8.8.9   *Mechanical and Electrical Safety*

Although mechanical and electrical safety are the primary responsibility of installation and service personnel, nuclear medicine technologists should be cognizant of the possibility of a mechanical or electrical hazard and report it immediately to the appropriate individual.

8.9 Nursing Personnel

8.9.1 Introduction

Radiation is one of the many occupational risks to which nurses are exposed in a medical setting. Effective radiation safety practices will keep exposures to nursing personnel to a minimum. Nurses *shall* be aware of the radiation safety policies regarding their specific work assignments and *should* insist that they be followed carefully, both by themselves and by other health care professionals. Nursing personnel frequently are present while patients receive diagnostic radiologic examinations and some are required to care for patients receiving radiation therapy with internally deposited or implanted radionuclides. These situations may offer some potential for exposure to radiation, and frequently the nurses are monitored individually by personnel dosimeters, either continuously or whenever patients containing therapeutic quantities of radionuclides are under their care.

8.9.2 Educational Requirements

Implementation of appropriate protective measures will maintain exposures of nurses at levels well below dose equivalent limits for radiation workers. It is essential that nurses have a working knowledge of these measures (see Section 7). Procedural guidelines for working with radiation sources *should* be a part of the nursing manual and should be readily available. Nursing personnel *should* receive periodic instruction regarding potential radiation sources and appropriate protective measures. If film badges or other dosimeters are issued to nursing personnel, instruction *should* be given on their use and on the implications of, and significance of, any readings received. If personnel dosimeters are not issued, the Radiation Safety Officer (RSO) *should* explain the justification for this decision during these instruction sessions.

The specific procedures outlined below can serve as guidelines for nursing practice, but the hospital RSO must be consulted for unusual situations or when there is concern for staff and patient safety.

8.9.3 Diagnostic X-ray Procedures

The question frequently arises as to who should hold a patient during a diagnostic x-ray examination. No individual medical employee *should* be assigned routinely to hold patients during diagnostic radiology procedures. Patients *should* be held only after it is determined that available restraining devices are inadequate. Individuals holding patients for x-ray procedures *should* be provided with lead aprons and lead gloves and *should* be positioned so that no part of the body is exposed to the direct radiation beam.

Diagnostic x-ray procedures, including some specialized procedures (Jankowski, 1984) such as cystoscopy, cardiac catheterization, angiography, and operating room procedures, require the presence of nursing personnel in proximity to patients for an extended time. The required protective measures for these nurses are identical to those followed by the radiologic technologist (see Section 8.4).

The majority of nursing personnel will be involved with x rays only while assisting in radiographic examinations using mobile equipment. Reasonable protection for them will normally be maintained during these procedures provided they:

- Do not permit themselves to be exposed to the direct beam.
- Remain at least 2m (6 ft) from the x-ray beam.
- Wear a leaded apron and gloves when they hold a patient or when it is necessary for them to remain closer than 2m (6 ft.) from the beam. (See NCRP, 1981 for requirements in the neonatal intensive care nursing).
- Hold patients only infrequently.
- *Should not* hold patients if they, the nurses, are pregnant.
- Follow any radiation safety instructions given.

8.9.4 *Diagnostic Nuclear Medicine Studies*

Patients undergoing nuclear medicine studies receive a small amount of a radionuclide, and therefore become potential sources of exposure to nurses (Burks *et al.*, 1982). The radionuclide used, its activity level, and any special precautions *should* be identified in the patient's chart. Several factors influence the exposure rate around the patient; the amount of radioactivity (becquerels or curies), the energy of the radiation, the physical half-life of the radionuclide (which determines the length of time during which the patient remains a significant source of radiation), the size of the patient (large patients absorb more of the gamma rays), and the rate of excretion or metabolism of the radiopharmaceutical. For example, technetium-99m (^{99m}Tc) utilized for bone scans is excreted rapidly in urine and frequent emptying of the bladder will lessen potential exposure to attending personnel as well as to the patient's gonads. Radiation measurements near patients indicate that remaining at a distance of several feet from the patient will normally provide adequate protection.

Many nuclear medicine scans utilize ^{99m}Tc with a physical half-life of six hours; exposure rates would be reduced by 50 percent in six hours due to physical decay alone. Very close patient contact over an extended time could result in a measurable dose, but documented radiation doses for nurses caring for these patients have not indicated any measurable dose. Since nursing care of these patients seldom requires very close proximity to the patient for more than a few minutes during any one hour, doses should be minimal. Care of these patients need not be restricted for pregnant nurses. Also, because of reduction of exposure with distance, exposure to other patients sharing a room would be minimal.

When the radionuclides used in nuclear medicine are injected into the patient or ingested, they mix with the patient's body fluids and excreta. It is therefore advisable for nurses to wear disposable gloves when handling these materials. For radionuclides having a half-life measured in days [for example, gallium-67 (^{67}Ga)], it may be necessary to use gloves for several days. The RSO *should* provide instructions for glove requirements and proper disposal. Excreta may be disposed of in the usual manner. A second flushing of centrally used bedpan hoppers is advisable. See Appendix B for more detailed information.

8.9.5 Therapeutic Radiation

Radiation is a common treatment for malignant disease. There are three main modes of treatment: (1) external radiation (*e.g.*, high energy x rays and electrons from linear accelerators), (2) permanent or temporary sealed (encapsulated) radionuclides and (3) unsealed radionuclides as solutions or colloidal suspensions.

Patients receiving external beam radiation therapy do not become a source of radiation exposure to persons around them. The patients themselves are irradiated when exposed to the primary beam of the machine, but following treatment, emit no radiation, just as following diagnostic x rays, patients emit no radiation. No radiation precautions are needed during subsequent nursing care, and patients may remain in close contact with family members.

The use of sources internally requires strict adherence to protective measures. Because the sources contain large quantities of radioactive materials, they expose the treated areas and surrounding healthy tissue and are a potential source of exposure to attending personnel (see Figure B.1, Appendix B). Detailed protective guidelines *should* be kept at the nurses station and a condensed form should be posted on the door of the patient's room. Pregnant nurses and aides *should* be restricted from providing care to these patients. (More detailed information concerning sealed and unsealed sources is provided in Appendix B).

Patient Cooperation

Prior to treatment with radionuclides, the patient *should* be given a careful explanation as to the nature of the treatment and the procedures involved. Patient cooperation is very important in minimizing unnecessary incidents and exposure. The need for restricting close contact time and limitations for visitors should be explained. Patients may feel very isolated with limited nursing care and restriction to their rooms; it is important to be caring and supportive of them.

Identification of Patient

There *shall* be a "CAUTION - RADIOACTIVE MATERIAL" label attached to the cover of, or in, the patient's chart. It *should* list the radionuclide and its activity (and date of determination or measurement), the exposure rate at 1 meter, and when precautions may be lifted. A similar tag *should* be attached to the patient, the patient's bed and on the door (NCRP, 1970b). Other hospital personnel (*e.g.*, dietary, aides, housekeeping) may enter the room briefly when permitted by and under the supervision of the head nurse.

Visitor Restrictions

Visitor restrictions are determined by the RSO and may be limited to one hour or less a day. Visitors are to remain as far as practical but at least six feet from the patient although momentary attendance at the bedside would be allowable. Pregnant women and young children *should not*, in general, be allowed to visit these patients.

Operating and Recovery Room Personnel

If radionuclides are implanted in the patient in the operating room, the same protective measures of time, distance, and shielding may be needed to protect operating room and recovery room personnel.

Personnel dosimeters may be issued to those working directly with the patient. Maintaining distance, as feasible, and efficient use of time are the primary means of minimizing personnel exposures. The RSO or designee may determine that lead shielding *should* be in place around the patient. In the recovery room, consideration *should* be given to the exposure of other patients. The patient with implanted radioactive source(s) *should* be located (in consultation with the RSO) so that no other patient receives an appreciable exposure, and reasonable efforts *should* be made to keep exposure to other patients as low as possible.

Emergency Procedures

Nurses may need to respond to emergencies relating to patients who have received therapeutic amounts of radioactive materials. Life-saving procedures *should* always begin as soon as possible without concern for exposure to radiation. The attending physician, radiation oncologist and RSO *should* be notified as soon as possible. For emergencies that are not immediately life threatening, the advice of the RSO *should* be sought on methods of reducing exposure. It is highly unlikely that a patient would contain a source of sufficient strength to be a significant health hazard to staff offering close emergency care.

Patient Death

In the event a patient dies after receiving treatment with unsealed radionuclides or while containing implants, the RSO *should* be notified. Tags identifying such patients as being radioactive *shall* be transferred to the outside of the shroud. Temporary implants *shall* be removed prior to transfer of the body.

8.9.6 *Summary*

Misunderstandings about the hazards of radiation may lead to unfounded concern for one's safety. Many medical professionals, ready to offer advice, are not fully knowledgeable about appropriate radiation protection. In-service programs on radiation *should* be scheduled annually for all nursing personnel and they *should* be encouraged to consult the RSO whenever there are concerns about improper procedures or their own personal exposures. In addition, the nursing service *shall* be represented on the hospital's Radiation Safety Committee; this offers an opportunity to voice nursing concerns not always readily acknowledged or recognized by the medical staff. It is essential that nurses, working in the presence of radiation, feel confident that sufficient rules and procedures are in place to offer protection equal to that provided for other hospital-related occupational hazards.

8.10 Pathologists/Morticians

8.10.1 Introduction

The contact of pathology personnel with radioactive materials falls into two categories: (a) laboratory testing and (b) post-mortem examination. Laboratory testing is discussed in Section 8.3. With regard to post-mortem examinations, there are generally three ways in which personnel can come in contact with radioactive materials: (1) death of a patient being treated with therapeutic amounts of radionuclides; (2) death of a patient given radionuclides for diagnostic purposes; or (3) handling specimens taken from these patients. Although radiation therapy in the form of brachytherapy is not usually administered to dying patients, there is the occasional death of such patients during treatment. Report No. 37 (NCRP, 1970b) covers this subject in detail and only a précis of that material is presented here.

8.10.2 Signs

Nursing personnel *shall* be responsible for the transfer of tags to the outside of the shroud identifying deceased patients as radioactive. Upon receipt of such bodies in the morgue, the Radiation Safety Officer (RSO) *should* be consulted for specific instructions.

8.10.3 Hazard Reduction

It should be possible to excise the area containing implanted sealed sources and to remove these sources to another table or area so that the post-mortem examination can proceed with less radiation exposure. Examination of the remaining specimens containing the sealed sources can be done under radiation protection control at some future time when the amount of radioactivity present will be less or when radiation safety personnel are on the scene to advise on desirable methods of radiation protection. The unwanted tissue specimens *should* be treated as radioactive waste.

8.10.4 Source and Fluid Removal

With liquid sources, specifically iodine-131 (^{131}I) or phosphorus-32 (^{32}P), body fluids may have to be considered contaminated and procedures supervised by the RSO. Double gloves *should* be worn to reduce the dose to the hands from beta particles. The external surface of the body and all portions of the autopsy table in contact with body fluids *should* be carefully washed down to remove any residual contamination. The table *should* be surveyed by the RSO to substantiate decontamination efforts.

With the use of technetium-99m (^{99m}Tc) labeled radiopharmaceuticals, the probability of patients expiring while still containing radioactivity is small. However, patients who expire within a few days after other nuclear medicine procedures, such as those with gallium-67 (^{67}Ga), may contain some residual radioactivity. The amount of radioactivity and consequent radiation exposure are not usually sufficient to be of special concern; however, the RSO *should* be contacted for guidance.

8.10.5 *Specimens*

Tissue sections from patients may be sent to pathology laboratory personnel for analysis and specimen preparation. Since such personnel are not usually familiar with techniques for handling radioactive materials, it is important that any such specimen be properly marked noting that radioactive materials are contained within. A separate section of the laboratory procedure manual dealing with radioactive specimens *should* be developed in consultation with the RSO.

8.11 Physicians (Non-Radiation Specialists)

8.11.1 *Introduction*

There are times when a physician may have under his care patients who are undergoing diagnosis or treatment with radiation sources. For this reason, the physician should have a general understanding of the characteristics of the radiation source and the associated safety procedures. Sections 1 to 7 of this report provide this basic information. The interested physician might also find continuing education courses on this subject at local and national meetings of professional societies.

8.11.2 *Authority/Responsibility*

When inpatients have been treated with sealed source implants or with therapeutic amounts of radionuclides, certain precautions must be exercised to reduce exposure of personnel who care for the patient. It is the responsibility of the physician in charge to see that instructions provided by the Departments of Radiation Therapy or Nuclear Medicine or the Radiation Safety Officer (RSO) are followed carefully.

8.11.3 *Emergency Response*

Physicians may need to respond to emergencies relating to patients who have received therapeutic amounts of radioactive materials. The amount of radioactive material used does not present a level of exposure to attending persons that would preclude an immediate response to a clinical emergency. Lifesaving procedures *should* always begin as soon as possible without concern for radioactivity. The attending physicians, radiation therapist and RSO *should* be notified as soon as possible. If the emergency is not immediately life threatening, the advice of the RSO *should* be sought for methods to reduce exposure.

8.11.4 *Patient/Family Relations*

While the patient is hospitalized, adherence to instructions that are provided will ensure that family members do not receive exposures beyond acceptable values. Once temporary sources have been removed, there is no further potential for exposure. When sealed sources cannot be removed, or when unsealed radionuclides are administered, the patient may continue to be a potential source of exposure to other people for some time, even after leaving the hospital. Specific instructions are usually given to the patient in writing by the physician responsible for administering the radionuclide or by the RSO.

8.11.5 *Radiation Accident Response Team*

The physician may be asked to assist or serve on the Radiation Accident Response Team because of his medical expertise. Any activities that might lead to exposure of the physician will be carefully monitored by the RSO or his designee.

8.12 Radiation Therapy Technologists

8.12.1 *Education*

Departments of radiation therapy *shall* provide an orientation program on radiation safety and *shall* further provide for the continuing education of the technology staff. Education should include attendance at professional meetings where innovations in treatment techniques and continuing education courses are presented. Inservice education should also be available for all technologists. At least once a year, the technology staff *should* meet with the Radiation Safety Officer (RSO) to review the radiation protection program and to discuss any problems which have been noted. These meetings *should* be documented. Continuing education *should* be recognized as essential to optimal patient treatment and staff safety.

8.12.2 *Monitoring*

Radiation therapy technologists *shall* be provided with personnel dosimeters. The technologist should be aware that the RSO will review the workload of the department and the types of procedures performed and may specify additional personnel monitors for extremities or for other critical organs. The technologist is responsible for using personnel monitoring devices as specified by the RSO or the Radiation Safety manual. The technologist *shall* report to the RSO immediately if personnel monitoring devices are lost, damaged or inadvertently exposed to radiation while not being worn. A technologist, if pregnant, should notify her supervisor and the RSO of the fact so that additional monitoring can be provided if necessary.

Upon installation of new radiation-producing equipment in a department, a survey *shall* be performed by a qualified medical or health physicist to verify that the radiation levels in the areas where the technologists will be working are within the guidelines or limits specified by the hospital policy and state and federal governments. The levels in these areas *shall* meet the requirements specified in Reports No. 49 and No. 51 (NCRP, 1976a; NCRP, 1977b) and *shall* be remeasured periodically for every permissible orientation of the beam. In some circumstances, it may be desirable to provide area monitors to verify the results of the survey under actual working conditions.

The technologist is responsible for taking full advantage of shielding provided to keep exposures as low as reasonably achievable and for using shielding devices as prescribed for each procedure.

8.12.3 Shielding

Shielding *should* be provided for all radiation-producing equipment in accordance with the recommendations specified in NCRP Reports No. 49 and No. 51 (NCRP,1976a, NCRP,1977b). Shielding requirements *should* be specified by a qualified medical or health physicist.

Shielding devices prescribed for use in treatment of patients *shall* be used and the technologist *should* ensure their use in the manner prescribed by the physician or physicist.

In departments employing brachytherapy, portable bedside shields may be used to reduce radiation exposure levels around patients containing implants.

8.12.4 Emergency Response

Technologists *should* be familiar with all potential emergencies involving radiation sources in their area. They also *should* be familiar with equipment and procedures that would be used in each emergency. Training and review programs *should* be conducted at least yearly and documented.

Emergency procedures for accidents involving patients *should* be posted at each treatment console. The technologist *should* be required to read these procedures, and a record acknowledging that this has been done *should* be maintained. The position of all emergency switches within the treatment room *should* be clearly identified by signs directly above or adjacent to the switch. Emergency OFF buttons shall be installed in the treatment room and on the console. A system "reset" *should* be provided at a separate location.

A radiation monitor, capable of indicating when the source is not completely within the source housing, or that the accelerator beam is on, *shall* be provided in the treatment room for teletherapy installations. This monitor *should* be independent of the teletherapy system and visible or audible to the technologist upon entering the treatment room.

The technologist *should* be familiar with systems which allow sources from teletherapy machines to be returned manually to shielded positions. These systems may vary among manufacturers. These systems *should* be tested as part of inservice education at least once a year.

8.12.5 *General Precautions*

The following conditions *shall* be followed for any radiation therapy equipment:

(a) Areas in which radiation therapy units are located *shall* be conspicuously posted with a sign bearing the radiation symbol and the words "CAUTION: HIGH RADIATION AREA". In addition, the controls *shall* be labeled "Caution: Radiation - This equipment produces radiation when energized".

(b) Radiation therapy units *shall* be operated only by designated personnel who have been approved by the Radiation Safety Committee or the RSO.

(c) Radiation therapy units *shall* be disconnected from their power source, or locked, when not in use.

(d) The operator *shall* ensure that only the patient is present and in the prescribed position before energizing the equipment. PATIENTS ARE NEVER HELD AND NO ONE IS PERMITTED IN THE TREATMENT ROOM DURING EXTERNAL BEAM RADIATION THERAPY. The operator *shall* be able to see and converse with the patient at all times during treatment. This is accomplished by the use of television, shielded windows, or mirrors, and a two-way intercom system.

(e) Appropriate personnel monitoring devices *shall* be worn by the operator and any assistants at all times.

(f) All personnel *shall* observe the procedures recommended by the RSO on the proper use of the radiation producing units.

(g) A door interlock *shall* be provided that allows a "Beam-ON" condition only when the door is closed.

(h) Independent back-up "Beam-ON" caution lights *shall* be provided on the console, above the treatment room door, and inside the treatment room.

(i) Emergency OFF buttons *shall* be installed in the treatment room and on the console.

8.13   Security Personnel

8.13.1   *Introduction*

Security personnel, like maintenance personnel, play an important role in the day-to-day operation of the radiation protection program. Typically, security personnel are on duty around-the-clock and are available after hours when anything unusual occurs. Security personnel should be aware of any unusual events or situations involving radiation sources. Such situations should be brought to the attention of the Radiation Safety Officer (RSO) immediately. As in the case of maintenance personnel, security personnel *should* maintain a listing of all areas in which radioactive materials are used or stored and a listing of the individuals to contact in an emergency. Such listings *should* contain succinct safety instructions for security personnel who have to enter such areas during emergencies when supervisors or the RSO may not be immediately available. Untrained security personnel should enter radiation areas only for life-saving situations or to prevent major damage to the facility, and time spent in the area should be minimal.

8.13.2   *Labels, Signs and Warning Lights*

Security personnel *should* be aware of the various radiation safety warning signs and labels. After hours, they *should* be on the alert for any such signs that may be found on items or receptacles in unusual places such as in the hallways. Security personnel *should* be aware that all areas in which radioactive materials are used or stored *should* be locked when the area is not occupied. In making their rounds they *should* check to ensure that all unoccupied radioisotope laboratories are locked and *should* report any discrepancies to the RSO.

8.13.3   *Response to Hazards or Accidents*

In the event of an unusual occurrence such as a fire, explosion or plumbing leakage in areas where radionuclides may be present, or when there is a radionuclide spill or any other type of contaminating event, security personnel *shall* keep all non-essential personnel out of the area and ensure that appropriate supervisory or radiation safety personnel have access to the area. In such situations, security personnel may be called upon to cordon off the area to keep unwanted people out or, in some cases, to keep people in. Security personnel *shall* ensure that all people in the area of an accident are removed from any life threatening situations and are appropriately monitored by the RSO before being allowed to leave the area. After hours, security personnel have the responsibility to notify the RSO of any such events and request specific precautions to be taken.

8.13.4 *Internal Receipt and Transport of Radioactive Materials*

Security personnel are usually assigned the responsibility of ensuring that packages of radioactive materials that are delivered to the hospital are delivered directly to the appropriate and designated receiving area or personnel. Security personnel should also be aware of the specific procedures for shipping and receiving of radionuclides. Night, weekend and holiday deliveries will be handled usually by security personnel. Upon arrival such deliveries *should* be taken immediately to the designated receiving site by the carrier who *should* be accompanied by a security officer. The security officer *should* visually inspect the package for any obvious signs of shipping damage and should immediately notify the RSO if any such damage is evident. If damage is evident, the security officer *should* request the carrier to remain to be monitored by the RSO . All packages of radioactive materials received after hours *should* be stored in a designated and locked area. All packages containing Yellow II or III shipping labels should be transported on a cart to maximize the distance between the transporter and the package.

8.14   Shipping and Receiving Personnel

8.14.1   *Introduction*

Specific regulations apply to the shipping or receiving of radioactive materials. Shipping and receiving personnel *should* be aware of these regulations and *should* receive periodic training from the Radiation Safety Officer (RSO).

8.14.2   *Receipt of Radioactive Materials During Normal Working Hours*

Immediately upon receipt of any packages of radioactive materials, the packages *should* be visually inspected for any signs of shipping damage such as crushed or punctured containers or signs of dampness. Any obvious damage *should* be reported to the RSO immediately. Do not touch any package suspected of leaking. Strongly urge the person delivering the package to remain until monitored by the RSO.

Since certain packages of radioactive materials will have detectable external radiation, they *should* be sent immediately to the designated check-in point (*e.g.*, the Radiation Safety Department or the Nuclear Medicine "Hot Lab") where they will be checked for contamination and external radiation level. They *should not* be allowed to remain in the receiving area any longer than is necessary as they may be a source of exposure for receiving personnel.

8.14.3   *Receipt of Radioactive Materials During Other Than Normal Working Hours*

Night, weekend and holiday deliveries will be handled usually by security personnel. Upon arrival, such deliveries *should* be taken immediately to the designated receiving site by the carrier who is accompanied by a security officer. The security officer *should* visually inspect the package for any obvious signs of shipping damage (crushed or punctured containers, wetness, etc.) and *should* immediately notify the RSO if any such damage is evident. If damage of the shipping container is evident, the security officer *should* request the carrier to remain for monitoring by the RSO.

All packages of radioactive materials received after hours *should* be stored in a designated and locked area.

All packages containing Yellow II or III shipping labels *should* be transported on a cart to maximize distance between the transporter and the package.

8.14.4 *Personnel Monitoring*

If shipping and receiving personnel follow the instructions listed above, there will be minimal chance for any detectable radiation exposure. The RSO *should* be consulted to review the situation and may issue personal dosimeters if appropriate.

8.14.5 *Shipping*

There are specific regulations covering the shipping of radioactive materials. The shipping department *should* accept for shipment only those packages of radioactive materials that have been prepared by or specifically approved by the RSO.

8.15   Ultrasonographers

Since ultrasound equipment does not produce or use ionizing radiation, ultrasonographers do not require personnel monitoring. However, since most ultrasound equipment is located in radiology departments and ultrasonographers may have to examine a patient shortly after a nuclear medicine study, the RSO may require the wearing of personnel dosimeters.

Ultrasonographers *should* contact the Radiation Safety Officer (RSO) for a review of their particular situation.

APPENDIX A

Spills

Accidental spillage of radioactive material is rare; however, spills may occur in the laboratory, in public areas such as the hall, the freight elevator, or in any hospital room or ward where a patient may vomit or be incontinent.

Major radiation accidents or serious spills of radioactive contamination have rarely involved medical and allied health personnel. Usually spills in hospitals have involved only small amounts of radioactivity in which a main concern is the spread of the contamination, *e.g.*, from shoes or contaminated clothing, into public areas. The following is a general outline of the procedure to be followed in the event of a spill.

1. Confine the spill immediately, by dropping paper towels or other absorbent material onto it.
2. Put on impermeable gloves.
3. Check shoes for visible signs of contamination. If it appears possible that they are contaminated, remove shoes when leaving the contaminated area.
4. Mark off or isolate in some way the entire suspect area and guard it to be sure that no one walks through it.
5. Detain all evacuees from the area in a place where they can be surveyed by the Radiation Safety Officer (RSO).
6. CALL THE RADIATION SAFETY OFFICER, (RSO). If the number is not posted in a convenient place, or you do not know it, call the telephone operator, report an emergency and request the RSO or the first accessible person on the radiation emergency call list.
7. In general, inexperienced personnel should not attempt to clean up a spill. It is better to wait a little while for the RSO than to risk spreading the contamination by erroneous pro-

cedures. If the spilled material is covered and bystanders are kept a few feet away, there is little or no danger from the radiation.

8. If any of the spilled material has splashed onto a person or clothing, immediate steps should be taken to remove it. Laboratory coats or outer garments should be removed and left in the contaminated area. Hands or other skin areas should be washed thoroughly with soap and water in the nearest wash basin, if by doing so the area of contamination is not enlarged. Care should be taken not to abrade or inflame the skin surfaces. If it is uncertain as to whether or not shoes are contaminated, the walkway to a washing facility shall be treated as a contaminated area until the RSO has certified that it is uncontaminated.

9. The RSO will bring decontamination materials and a survey meter and the cleanup operation will proceed.

10. If the RSO is not immediately available and cleanup must be initiated, as few people as possible should be involved in the actual decontamination efforts. Impermeable gloves, shoe covers and a surgical face mask should be worn if available. The spilled material shall be taken up with absorbent paper, which is handled with forceps or tongs, and deposited immediately in a waterproof container. After as much contamination as possible has been removed in this way, the surface should be washed with damp - not wet - paper towels held in forceps, always working toward the center of the contaminated area rather than away from it.

11. A survey meter should be available, and careful monitoring of both area and personnel should be carried out during this procedure. The survey meter should be operated by someone who is not involved in the cleanup, so that the instrument does not become contaminated. Cover the probe with thin, clear plastic wrap, if possible.

12. Reduction of the counting rate to several times background is usually satisfactory. Higher counting rate areas can be covered with plastic-backed absorbent paper and held in place with tape to await further evaluation by the RSO. The RSO should survey the area and certify adequate decontamination prior to its return to routine use.

13. When the operation is finished, gloves and other protective garments should be checked carefully for residual contamination. If any is found, the garments should be left with the other contaminated material in plastic bags for ultimate disposal by the RSO.

14. Life saving efforts and vital first aid have priority over contamination concerns.
15. If necessary, activate the medical radiation emergency plan.

Loss of a Sealed Source

Immediately upon discovery of a loss of a sealed source, an appropriate plan of action should be initiated. An example of such a plan would be as follows.

1. Call the Radiation Safety Officer immediately.
2. Make a list of all possible places in which the source might have been and where it might be found.
3. Choose the most sensitive and appropriate portable survey instruments (*e.g.*, μR meters or portable scintillation detectors for gamma or high energy beta emitters) for conducting the search.
4. If the source had been transported, check the entire route of travel.
5. If the source had been used with a patient, survey the patient, the patient's room and all bandages, linen, bedding and trash from the patient's room.
6. Survey the entire route from the patients room to the laundry and the laundry facility.
7. Survey the entire route from the patient's room to the incinerator, the incinerator, trash awaiting incineration and the incinerator ash.
8. Survey the entire route from the patients' room to the dumpster and the trash in the dumpster. If needed, request Security to impound the dumpster until the search can be completed.
9. If instruments had been used with the patient, survey the entire route from the patients' room to the instrument cleaning and sterilization area.
10. Survey all areas where the source might be found, such as sink drains or plumbing fixtures, elevator shafts, waste cans, trash bins and vacuum cleaners or house vacuum systems.
11. Continue the search until the source is found or the search is terminated by the RSO.

A Ruptured or Broken Sealed Source

1. Shut off all fans and ventilators.
2. Drop damp towels on the suspect material; throw nothing away.

3. If possible, evacuate the room. If not, keep all personnel as far as possible from the suspect material until the RSO arrives. Do not attempt to clean up or remove any material.
4. Call the RSO to remove the questionable material and check the area for contamination.

APPENDIX B

Special Considerations for Patients Containing Sealed or Unsealed Therapy Sources

Introduction

The effectiveness of radiation therapy is determined partially by the relative radiosensitivity of the treated cells and the amount of energy deposited into those cells. A positive therapeutic effect is dependent upon optimizing the differential deposition of energy in diseased cells versus normal cells. Radionuclides serve as a concentrated source of radiation which deposit their energy locally to maximize the therapeutic effect without damaging adjacent healthy tissue.

Therapeutic procedures using radionuclides and the precautions associated with these procedures are dependent upon the disease site and the mode of therapy designed for the treatment. Procedures involving sealed sources are different from procedures involving unsealed sources, and intracavitary insertions require different considerations from interstitial placement of the radioactive material. However, there are some general considerations to be followed for all radionuclide therapy procedures.

General Considerations

1. It is important for the patient to understand the nature of the treatment. Patient cooperation is important in minimizing unnecessary incidents and exposure.
2. Prior to the administration of the radionuclide, the procedures and special precautions should be reviewed with the nursing staff. The nursing staff *shall* have specific written instructions for each procedure and *should* review them before the patient arrives in the room.
3. Immediately following the return of the patient to the hospital room, or after the administration of the compound or insertion of the sources, a person from the Radiation Safety Office *should*

survey the patient and surrounding areas to determine distance and time restrictions for hospital personnel and visitors in the patient's room. These distances and times are recorded on a form in the patient's chart and listed on the caution sign on the patient's door. These signs and labels *should* remain posted until removal is ordered by the RSO.

4. Hospital personnel and allowed visitors *should* position themselves as far from the patient as is reasonable except for necessary bedside care. A distance of 2 meters is normally acceptable. In some cases, the RSO may determine that mobile lead shields are needed to reduce exposure to others in adjacent areas. Specific restrictions will be noted by the RSO on the room door and in the hospital chart.

5. It is not advisable for pregnant women or children under age 18 to enter the hospital room.

6. Personnel dosimeters are required for all hospital personnel who are likely to receive in excess of 25 percent of the dose equivalent limit for radiation workers. The RSO will identify hospital personnel within this category and issue the appropriate dosimeters to them.

7. Pregnant personnel *should not* routinely be assigned to the care of patients under treatment with radioactive materials.

8. Patients receiving radionuclide therapy *should* be assigned a private room and should be restricted to the room unless an exception is authorized by the RSO.

Specific Considerations.

1. *Brachytherapy*

Most patients who receive intracavitary brachytherapy are being treated for gynecological cancer. The procedure requires the temporary (24-48 hour) insertion of an applicator into the uterus and/or vaginal vault. The applicator is inserted into the patient in the operating room, and the position of the applicator is verified radiographically. Since no radiation source is present at this point in the procedure, the only radiation exposure to operating room personnel is from the x-ray machine. After radiographic verification, the radioactive sources are inserted into the applicator, usually in the patient's room, but sometimes in another area of the hospital. In the latter case, the patient is transported to the room after insertion of the sources. In either case, the patient becomes a source of exposure to

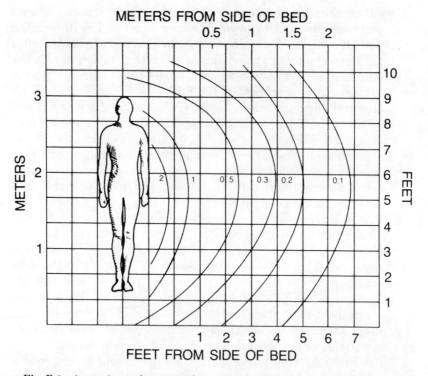

Fig. B.1 Approximate dose equivalent rates in mSv h⁻¹ (mrem h⁻¹ x 100) in regions around the bed of a patient containing 3700 MBq (100 mCi) cesium-137 or 5550 MBq (150 mCi) of iodine-131 (NCRP, 1976c). Note that although only one side is shown, the exposure pattern completely surrounds the body.

others as soon as the sources are placed in the applicator, and thus, precautions are initiated at that time. Exposure rates around the patient will depend on the total activity of the radionuclide source and the patient thickness. Nominal values are indicated in Figure B.1. The following instructions are for hospital personnel involved in the care of the patient with an intracavitary radionuclide insertion:

1. If it is necessary to move the patient after the sources have been inserted, the transporter *should* grip the gurney as far from the sources as possible. The transporter *should* proceed directly to the destination using a restricted elevator and *should* avoid unnecessary proximity to other patients or personnel. Personnel monitoring requirements *shall* be assessed by the RSO.

2. All personnel including residents, fellows, transporters and nurses *should* be familiar with the appearance of the applicator

and sources used at their institution. If any one suspects an unusual circumstance regarding the applicator or the sources, their supervisor and the RSO *should* be notified immediately.

3. Needles, capsules, or containers holding brachytherapy sources *shall* not be handled directly. If a source becomes dislodged, long forceps *should* be used to put the source in the corner of the room farthest from the corridor in the shielded container provided. The RSO and the Radiation Therapy Department *should* be notified immediately.

4. Bed baths given by the nurse *should* be omitted while the sources are in place.

5. Perineal care should not be given during gynecologic treatment; the perineal pad may be changed when necessary unless orders to the contrary have been written. Perineal pads are not contaminated.

6. No special radiation safety precautions are needed for sputum, urine, vomitus, stools, dishes, instruments, or utensils unless specifically ordered.

7. All bed linens *shall* be checked with a radiation survey meter by radiation safety personnel before being removed from the patient's room to ensure that no dislodged sources are removed inadvertently.

8. At the conclusion of treatment, (1) call the RSO to survey the patient and room; (2) radiation therapist or radiation safety personnel *should* count the radiation sources to be sure that all temporary implants have been removed prior to discharging the patient and return the sources to the storage vault.

2. Patients Treated with Radioactive Iodine

Patients receiving iodine-131 (^{131}I) treatments for thyroid disorders require special precautions (Miller *et al.*, 1979) in addition to the general considerations given above to prevent contamination and spread of radioactivity outside the restricted treatment area. The treatment of thyroid cancer requires the use of relatively large amounts of ^{131}I, and all body fluids such as saliva, sweat, urine, feces and vomitus will contain radioactivity. Radiation protection of personnel caring for these patients requires the cooperative effort of the patient, nursing staff, the RSO and all individuals who interact with the patient. The following general guidelines apply to the care of these patients.

1. Disposable gloves and shoe covers *should* be positioned outside the patient's room, and a clearly labelled waste receptacle

should be positioned just inside the door of the room to avoid transport of contamination to other areas. This will enable necessary hospital personnel to protect themselves from contamination when they enter the patient's room. Pocket dosimeters *should* be available for estimating the exposure of these personnel.

2. Urine and vomitus from ^{131}I therapy patients may be disposed of by way of the sewer or stored for decay in the radioactive waste storage area. The method of disposal *should* be determined by the RSO.

3. Patients containing ^{131}I *shall* be confined to their rooms except for special medical or nursing purposes approved by the Nuclear Medicine or Radiation Therapy Department and the RSO.

4. Attending personnel *should* wear rubber or disposable plastic gloves when handling urinals, bedpans, emesis basins, or other containers having any material obtained from the body of the patient. Gloves *should* be washed before removing them and then the hands *should* be washed. The gloves *should* be left in the patient's room in the designated waste container. These gloves need not be sterile or surgical in type.

5. Disposable items *should* be used in the care of these patients, whenever possible. After use, these items *should* be placed in the designated waste container. Contact the RSO for proper disposal of the contents of this designated waste container.

6. All clothes and bed linens used by the patient *should* be placed in the laundry bag provided and *should* be left in the patient's room to be checked by the RSO.

7. All nondisposable items *should* be placed in a plastic bag and *should* be left in the patient's room to be checked by the RSO.

8. If there are surgical dressings, they *should* be changed only as directed by the physician. Dressings *should* not be discarded but *should* be collected in plastic bags and turned over to the RSO. Handle these dressings only with tongs or tweezers. Wear disposable gloves.

9. If the urine from ^{131}I patients is to be collected, special containers *should* be provided by the RSO. The patient *should* be encouraged to collect his own urine in the container. If the patient is bedridden, a separate urinal or bed pan *should* be provided. The urinal or bed pan *should* be flushed several times with hot soapy water after each use.

10. Disposable plates, cups, and eating utensils *should* be used by patients who are treated with ^{131}I.

11. Vomiting within 24 hours after oral administration, urinary continence, or excessive sweating within the first 48 hours

may result in contamination of the linen and floor. If any such situation occurs, or if radioactive urine and/or feces is spilled during collection, call the RSO. Meanwhile, attending personnel *should* wear shoe covers and handle all contaminated material with disposable gloves and avoid spreading contamination.

12. All vomitus *should* be kept in the patient's room for disposal by the RSO. Feces need not be saved routinely, unless so ordered on the chart. The same toilet *should* be used by the patient at all times, and it *should* be flushed several times after each use.

13. Precautions should be taken to see that no urine or vomitus is spilled on the floor or the bed. If any part of the patient's room is suspected of being contaminated, notify the RSO.

14. When exiting the room, be sure to remove shoe covers at the doorway. A step-off pad placed at the doorway will help to ensure stepping out into an uncontaminated area once the shoe covers have been removed.

15. If a nurse, attendant, or anyone else knows or suspects that his or her skin or clothing, including shoes, is contaminated, the RSO *should* be notified immediately. This person *should* remain in an area adjacent to the patient's room and *should* not walk about the hospital. If hands become contaminated, wash them immediately with soap and water.

16. If a therapy patient should need emergency surgery or should die, the RSO and the Nuclear Medicine or Radiation Therapy Department *should* be notified immediately.

17. Patient discharge *shall* be recommended by the RSO. Generally the patient must contain less than 1100 MBq (30 mCi) ^{131}I (NRC, 1987) before being discharged. The dose equivalent rate at one meter from the patient should be less than 0.05 mSv h^{-1} (5 mrem h^{-1}). Following discharge of the patient, the RSO *should* have the room surveyed and *should* supervise decontamination procedures before the room is returned to routine use.

18. Before a therapy patient's room is reassigned to another patient, the room *should* be surveyed for contamination and decontaminated if necessary, and all radioactive waste and waste containers *should* be removed.

APPENDIX C

Definitions

absorbed dose (*D*): The energy imparted to matter by ionizing radiation per unit mass of irradiated material at the point of interest; unit of absorbed dose has been the rad and now in the System International (SI) is the gray (Gy), 100 rad = 1 Gy.

absorption: Transfer of energy from ionizing radiation to an absorber (*e.g.*, tissue); see attenuation.

accelerator: A device that accelerates charged particles (*e.g.*, protons or electrons) to high speed, often used for the production of certain radionuclides or for treatment of radiation therapy patients.

activity: The number of nuclear transformations per unit time; units of activity are becquerel or curie.

air kerma: see kerma.

ALARA (As Low As Reasonably Achievable): The principle of limiting the radiation dose of exposed persons to levels as low as is reasonably achievable, economic and social factors being taken into account.

alpha radiation (or particles): Charged particles consisting of two protons and two neutrons in close, stable association (Helium nucleus), emitted from a radioactive nucleus during decay.

attenuation: Loss of energy from ionizing radiation by scatter and absorption.

background: Ionizing radiation present in the region of interest and coming from sources other than that of primary concern. (See Natural Background Radiation also).

becquerel (Bq): The special name for the unit of activity in SI. 1 Bq = 1 s^{-1}, *i.e.*, one disintegration per second. 1 Bq = 27 x 10^{-12} Ci.

beta radiation (or particles): Electrons emitted from a radioactive nucleus during decay.

brachytherapy: Use of an encapsulated source to deliver gamma or beta radiation at a distance up to a few centimeters by surface, intracavitary or interstitial application.

contamination: Radioactive material present in undesired locations, particularly where its presence may be harmful.

collective dose equivalent: The sum of the individual dose equivalents received in a given period of time by a specified population from exposure to a specified source of ionizing radiation.

curie: The conventional unit of activity of radioactive material decaying at the rate of 3.7×10^{10} transformations per second (roughly equivalent to the activity of 1 gram of radium). See becquerel. $1 \text{ Ci} = 3.7 \times 10^{10} \text{ Bq}$.

dose equivalent (H): A quantity used for radiation protection purposes that expresses on a common scale for all radiations, the irradiation incurred by exposed persons. It is defined as the product of the absorbed dose, D, and the quality factor, Q. The name for the unit of dose equivalent (J kg^{-1}) is the sievert, Sv. 1 sievert = 100 rem. (See sievert, rem).

dose equivalent limit (annual) (H_L): The annual dose equivalent limit, H_L, defines the degree to which radiation exposure should be controlled to achieve an acceptable level of risk for workers and the general public, taking into account both somatic and genetic detriment. The present limit for annual occupational exposure is 50 mSv (5 rem) and the corresponding limit for exposure to the public is 1 mSv (0.1 rem) for continuous exposures and 5 mSv (0.5 rem) for infrequent exposures.

dosimetry: The science or technique of determining radiation dose.

dosimeter: Dose measuring device. See also personnel dosimeter.

early somatic effects: Radiation effects occurring shortly after exposure to high doses of radiation; these include erythema, epilation, and anorexia.

effective dose equivalent (H_E): The sum, over specified tissues, of the products of the dose equivalent in a tissue (T) and the weighting factors for that tissue, (w_T), i.e., $H_E = \Sigma \, w_T \, H_T$.

exposure: The incidence of ionizing radiation on living or inanimate material. Also, a measure of the ionization produced in a specified mass of air by x or gamma radiation, which may be used as a measure of the ionizing radiation to which one is exposed. When using SI units, air kerma is often used in place of exposure. Air kerma has the units of J kg^{-1} (gray). In conventional units, the special unit of exposure is the roentgen, R. An exposure of 1 R corresponds to an air kerma of 8.7 mGy. (See kerma, gray, roentgen).

film badge: An assembly containing a packet of unexposed photographic film and a variety of filters (absorbers); when the film is developed, the dose and type of radiation to which the wearer was exposed can be estimated.

gamma rays: Electromagnetic radiation emitted in the process of nuclear transition.

gauss: A unit of magnetic induction. The earth's magnetic field is approximately 0.5 gauss.

Geiger-Mueller (GM) counter: Highly sensitive, gas-filled radiation-detecting device, which is capable of detecting individual nuclear particles.

genetic effects: Changes in reproductive cells that may result in abnormal offspring of persons or animals.

gray (Gy): The special SI unit of absorbed dose and kerma equal to 1 J kg^{-1} in any medium. 1 Gy = 100 rad.

half-life: The time required to reduce the amount of a radionuclide to one half the amount originally present. Physical or radioactive half-life refers to reduction of activity by radioactive decay; biological half-life refers to biological elimination from the body; and effective half-life refers to the combined action of radioactive decay and biological elimination.

in utero: In the uterus; refers to a fetus or embryo.

in vitro: Refers to a procedure carried out in a test tube or other vessel, *e.g.,* tissue culture study: See *in vivo.*

in vivo: Refers to a procedure carried out in the living body: See *in vitro.*

inverse square law: A physical law stating that the intensity of x or gamma radiation from a point source emitting uniformly in all directions is inversely proportional to the square of the distance from the source. Example: A point source that produces 10 Gy/h at 1 m will produce 2.5 Gy/h at 2 m.

ionization chamber: A device for detection of ionizing radiation or for measurement of radiation dose and dose rate.

ionizing radiation: Electromagnetic radiation (x or gamma rays) or particulate radiation (alpha particles, beta particles, electrons, positrons, protons, neutrons and heavy particles) capable of producing ions by direct or secondary processes in passage through matter.

isotope: In the strict sense, any of two or more species of a chemical element with the same atomic number but different atomic weight; often incorrectly applied to radionuclide or radiopharmaceutical.

kerma (K): Kerma (Kinetic energy released in material) is a quantity that represents the kinetic energy transferred to charged particles by uncharged particles per unit mass of the material irradiated. The SI unit for kerma is the J kg^{-1} with the special name of gray, Gy. 1 Gy 100 rad (See gray).

late somatic effects: Radiation effects that may occur in individuals a considerable time after exposure to radiation; these include mutagenic effects and carcinogenic effects.

monitor: To determine the level of ionizing radiation or radioactive contamination in a given region. Also, a device used for this purpose.

natural background radiation: Radiation originating in natural sources: for example, cosmic rays, naturally occurring radioactive minerals, naturally occurring radioactive carbon-14 and potassium-40 in the body.

nonstochastic effects: Effects, for which the severity of the effect in affected individuals varies with the dose, and for which a threshold exists.

nuclide: A species of atom characterized by its atomic number, mass number, and nuclear energy state, provided that the mean life in that state is long enough to be observable.

occupationally exposed: Exposed to radiation as a direct result of occupational duties.

personnel dosimeter: A small radiation detector that is worn by an individual. Common types include film badges, TLD badges, pocket dosimeters and pocket ionization chambers.

photon: A quantum of electromagnetic radiation.

qualified expert: A qualified expert, for radiation protection purposes, is a person having the knowledge and training to measure ionizing radiation, to evaluate safety techniques, and to advise regarding radiation protection needs (for example, persons certified in an appropriate field by the American Board of Radiology, or the American Board of Health Physics or the American Board of Nuclear Medicine Science, or persons otherwise determined to have equivalent qualifications).

rad: The conventional unit of absorbed dose equal to the absorbed energy of 0.01 J kg^{-1} (100 ergs g^{-1}) in any medium and is being replaced by the gray. 1 rad $= 0.01$ Gy. (See note under rem).

radio-: A general prefix relating to radiation (*e.g.*, radiosensitive) or state of being radioactive (*e.g.*, radioiodine).

radioactive waste: Waste with levels of radioactivity sufficient to exceed background levels, and to be of potential concern to health, thereby requiring special storage, transportation, and disposal methods.

radioisotope: An unstable atom having the same atomic number but a different number of neutrons in the nucleus than the comparable stable atom. (See isotope).

radionuclide: A radioactive nuclide. (See nuclide).

radiopharmaceutical: A radioactive drug product administered to a patient for diagnostic or therapeutic nuclear medicine procedures.

rem: The conventional unit of dose equivalent. The dose equivalent in rem is numerically equal to the absorbed dose, D, in rad multiplied by the quality factor Q. 1 rem = 0.01 sievert. (See sievert, rad). [Note: for most medical applications involving x ray or gamma emitters, the numerical values of the absorbed dose in rad, dose equivalent in rem, and exposure in R are roughly equivalent numerically].

roentgen (R): The special unit of exposure, based on a quantity of ionization (charge) produced by the absorption of x or gamma radiation energy in a specified mass of air under standard conditions. 1 R = 2.58 x 10^{-4} C kg^{-1} of air. For radiation protection purposes, an exposure to 1 roentgen of x or gamma rays (air kerma of ~10^{-2} J kg^{-1}) is generally assumed to produce an absorbed dose of 1 rad in water or soft tissue. (See exposure, gray, kerma).

scatter: Deflection of radiation passing through matter, causing change of direction of subatomic particles or photons, attenuation of the radiation beam and usually some absorption of energy.

sievert (Sv): The special name sievert (Sv) has been adopted for the SI unit of dose equivalent. The dose equivalent, H, is the product of D, and Q, at the point of interest in the tissue where D is the absorbed dose and Q is the quality factor. 1 Sv = 100 rem. (See rad, rem, roentgen).

somatic effects: Detrimental effects of radiation manifested in the person irradiated.

stochastic effects: Effects, the probability of which, rather than their severity, is a function of radiation dose without threshold. (More generally, stochastic means random in nature).

survey meter: An instrument or device, usually portable, for monitoring the level of radiation or of radioactive contamination in an area or location.

syringe shield: A cylinder made of lead or lead-containing glass that absorbs radiation emitted from the contents of a syringe, protecting the user.

thermoluminescent dosimeter (TLD): A dosimeter containing a crystalline solid for measuring radiation dose, plus filters (absorbers) to help characterize the types of radiation encountered. (When heated, TLD crystals that have been exposed to ionizing radiation give off light proportional to the energy they received from the radiation.)

wipe test: A test for radioactive contamination. A surface is wiped with a small paper or cloth which is then tested for radioactivity.

APPENDIX D

Sources of Nonionizing Radiation in Medical Facilities

D.1 Ultrasound

Although ultrasound is not a source of ionizing radiation, it is found in routine use in the radiology department and ultrasonographers may work with patients subsequent to a nuclear medicine study. Therefore, in some institutions, the RSO may require them to wear personnel dosimeters to determine if they receive exposure to ionizing radiation from these nuclear medicine patients.

The ultrasound frequency is higher than can be detected by humans. In medicine, the commonly used frequencies range from 1 to 10 MHz (1 to 10 million cycles per second). When high frequency sound is transmitted through a medium like water, or the body, objects intercepting these sound waves reflect them back to their origin. The amount of time it takes for these sound waves to travel back to the transmitter is directly related to their depth in the body. The use of high frequency sound waves to image organs is a commonly chosen alternative to the use of ionizing radiation in diagnostic procedures. Further information on ultrasound is available in NCRP Report No. 74 (NCRP, 1983b) and NCRP Report No. 73 (NCRP, 1983c).

D.2 Video Display Terminals

Many employees spend a considerable portion of their working day in front of a video display terminal (VDT). VDT circuits produce several forms of radiation including visible light, infrared radiation and x radiation. The image on the screen is created by electrons striking a fluorescent screen within the cathode ray tube. However, shielding has been incorporated in the tube-facing to filter out the low energy x rays produced, and exposure levels from nonionizing radiations are well below accepted occupational and environmental health and safety standards (FDA, 1981; Hirning and Aitken, 1982; ACGIH, 1983; NRC, 1983). Lighting, window glare, angle of viewing,

99

keyboard position and chair height are all factors that can influence the health and well being of the operator and *should* be considered.

It has been suggested that ionizing radiation emitted by VDT screens may cause increased incidence of miscarriages among VDT users. No scientific data support this contention. In fact, federally sponsored and privately funded studies with highly sensitive instrumentation have determined that ionizing radiation is not detectable from most VDT screens.

D.3 *Magnetic Resonance Imaging*

Unlike conventional radiographic procedures that utilize ionizing radiation, magnetic resonance imaging (MRI) produces an image of patient anatomy without x rays or gamma rays. MRI uses the differences in magnetic properties of protons in hydrogen atoms as a means of distinguishing among the different tissues in the patient. A magnetic field from approximately 500 to 15,000 gauss or more [the earth's field is approximately 0.5 gauss] may be used to produce images of the patient. In addition to the large static magnetic field, the patient is exposed to changing gradient magnetic and radiofrequency electromagnetic fields to localize and induce a signal. The technologists who operate the MRI equipment and assist the patient in entering and leaving the magnet are exposed to magnetic fields hundreds to thousands of times greater than the earth's magnetic field. The health effects of this level of magnetic field strength are considered to be small.

To date, the most significant hazard for technologists, housekeeping, and service personnel is the risk of projectile movement of ferrous metallic objects in the magnetic field. The fringe field near the magnet is strong enough to pull a loosely held pair of scissors or other object from the hand of an unsuspecting person. The magnetic force is strong enough to accelerate the object toward the magnet with enough velocity to harm an individual in its path.

The entrance to the scanner area *shall* be posted with signs alerting persons entering the area of the presence of the magnetic field.

There is some uncertainty regarding the effect of the magnetic field on some heart pacemakers. A prudent guideline is to exclude people with pacemakers from the area near the MRI unit. Problems may also be encountered in patients with aneurism clips. The decision to scan these patients must be left with the radiologist responsible for the examination.

Special care in the scanner area is required for patients on life support systems, or those requiring life support systems. In addition to the risk of metallic objects being drawn into the magnet, some equipment, *e.g.*, defibrillators, and resuscitation equipment, may not work properly in the vicinity of the magnet. The hospital *should* review and establish guidelines to be followed in the event emergency care is required in the scanner area.

D.4 *Radiofrequency Radiation*

Radiofrequency electromagnetic radiation (RFEM) is the portion of the electromagnetic spectrum with waves that range in frequency from >0 to 3×10^{12} Hz which is below the infrared portion of the spectrum. Many communication systems use RFEM radiation, and consequently the population is continuously exposed to this type of radiation. The primary mechanism of interaction of high levels of RFEM radiation with biological systems is the deposition of energy in the form of heat. The rise in temperature is proportional to the power deposition in the exposed tissue. Power deposition in excess of recommended limits should be avoided (ANSI, 1982; ACGIH, 1983; NCRP, 1986). In addition to heating effects, RFEM radiation is known to interfere with the operation of some pacemakers, electronic thermometers and other electronic devices. Specific information on the biological effects of RFEM including exposure criteria can be found in NCRP Report No. 86 (NCRP, 1986). That report also contains a discussion of the biological responses to weak RFEM fields, *i.e.*, those fields that do not cause significant increase in temperature (NCRP, 1986, Chapter 11).

Microwaves which are RFEM radiations at frequencies of 300 MHz to 300 GHz are found in the medical environment associated with various techniques and devices such as linear accelerators, hyperthermia and diathermy units and microwave ovens. Microwaves used in association with these units are produced only during operation. The FDA safety standard for microwave ovens limits the microwave emission to 5 mW/cm^2, measured at a distance of 5 cm from the external surface of the oven (FDA, 1986).

D.5 *Lasers*

Lasers produce very intense and collimated beams of light. Certain lasers used in medicine employ a wavelength that is ideal for burning or cutting through tissue in a very narrow slice. Such devices have been used for years in ophthalmology to perform surgical procedures on the eye. Newer lasers are increasingly being used in other areas of surgery.

Laser beams travel in straight lines until they are intercepted or reflected; therefore, they can cause damage at considerable distances (across rooms or in other rooms or corridors if transmitted through windows or open doorways). In surgical procedures, it is imperative that the laser be rigidly aligned before activation and firing. Otherwise, inadvertent damage can occur to the patient's healthy tissue, the user or attendants.

Because of the intense energy contained in the laser beam it can be a further hazard if allowed to impinge on flammable or explosive materials. See FDA, 1986 for a discussion of safety requirements placed on manufacturers of laser producing equipment. Lasers *shall* be used only by those individuals specifically trained and approved as users in writing by the appropriate safety committee. Such usages need to be controlled rigidly and conducted in accordance with specific protocols. Eye protection *shall* be provided where indicated and all specularly reflective materials covered or removed from the area in which lasers are being used. Access to the usage area *shall* be controlled when the laser is in operation.

Low-powered lasers are used for alignment of the patient in radiation therapy, CT scanning, magnetic resonance imaging and nuclear medicine. These lasers do not present the hazard noted above nor require the precautions.

References

ACGIH (1983). American Conference of Governmental Hygienists. *TLV's. Threshold Limit Values for Chemical Substances and Physical Agents in the Workplace for 1983-1984.* (American Conference of Governmental Hygienists, Cincinnati, Ohio).

ANSI (1979). American National Standards Institute. *Safety Color Code for Marking Physical Hazards,* ANSI Z53.1-1979 (American National Standards Institute, New York).

ANSI (1982). American National Standards Institute. *Safety Levels with Respect to Human Exposure to Radio Frequency Electromagnetic Fields, 300 kHz to 100 GHz,* Report No. ANSI C95.1-1982 (The Institute of Electrical and Electronics Engineers, New York).

BARRALL, R.C. AND SMITH, S.I. (1976). "Personnel radiation exposure and protection from 99mTc radiations," pages 77 to 97 in *Biophysical Aspects of the Medical Use of Technetium-99m,* Kereiakes, J.G. and Corey, K.R., Eds. AAPM Monography No. 1 (American Institute of Physics, New York).

BURKS, J., GRIFFITH, P., McCORMICK, K., AND MILLER, R. (1982). "Radiation exposure to nursing personnel from patients receiving diagnostic radionuclides," Heart Lung 11, 217-220.

CASTRONOVO, F.P., WEBSTER, E.W., STRAUSS, K.W., BREEN, C.. HOLLY, M. AND FOLDING, F. (1982). *A Health Physics Guide for Patient Care Units: The Radiation Precautions Associated with Patients Undergoing Diagnostic Radiopharmaceutical Procedure* (Massachusetts General Hospital, Boston).

FDA (1977). Food and Drug Administration. *The Mean Active Bone Marrow Dose to the Adult Population of the United States from Diagnostic Radiology* DHEW Publication (FDA) 77-8013 (Government Printing Office, Washington).

FDA (1981). Food and Drug Administration. *An Evaluation of Radiation Emission from Video Display Terminals,* HHS Publication (FDA) 81-8153 (Government Printing Office, Washington).

FDA (1986). Food and Drug Administration. Code of Federal Regulations. Title 21, Chapter 1, Subchapter J - Radiological Health, Parts 1000-1050 (Government Printing Office, Washington).

HIRNING, C.R. AND AITKEN, J.H., (1982). "Cathode-ray tube x-ray emission standard for video display terminals," Health Phys. 43, 727-731.

JANKOWSKI, C.B. (1984). "Radiation exposure of nurses in a coronary care unit," Heart Lung 13, 55-58.

KACZMAREK, R.G., BEDNAREK, D.R., WONG, R., KACZMAREK, R.V., RUDIN, S., AND ALKER, G. (1986). "Potential radiation hazards to personnel during dynamic CT," Radiology 161, 853 (Letter to the Editor).

LAND, C.E. (1980). "Estimating cancer risks from low doses of ionizing radiation", Science **209**, 1197-1203.

LIS, G.A., SU'BI, SAID, M. AND BRAHMOVAR, S.M. (1981). "Fingertip and whole body exposure to nuclear medicine personnel," J. Nucl. Med. Technol. **9**, 91-98.

MILLER, K.L., BOTT, S.M., VELKLEY, D.E. AND CUNNINGHAM, D.E. (1979). "A review of the contamination and exposure hazards associated with therapeutic uses of radioiodine," J. Nucl. Med. Technol. **7**, 163-166.

NAS (1980). National Academy of Science. National Research Council Committee on the Biological Effect of Ionizing Radiation, *The Effects on Populations of Exposure to Low Levels of Ionizing Radiation*, (BEIR III) (National Academy Press, Washington).

NAS (1988). National Academy of Sciences. National Research Council Committee on the Biological Effects of Radiation. *Health Risks of Radon and Other Internally Deposited Alpha-Emitters* (BEIR IV) (National Academy Press, Washington).

NCRP (1970a). National Council on Radiation Protection and Measurements. *Dental X-Ray Protection*, NCRP Report No. 35 (National Council on Radiation Protection and Measurements, Bethesda, Maryland).

NCRP (1970b). National Council on Radiation Protection and Measurements. *Precautions in the Management of Patients Who Have Received Therapeutic Amounts of Radionuclides*, NCRP Report No. 37 (National Council on Radiation Protection and Measurements, Bethesda, Maryland).

NCRP (1970c). National Council on Radiation Protection and Measurements. *Radiation Protection in Veterinary Medicine*, NCRP Report No. 36 (National Council on Radiation Protection and Measurements, Bethesda Maryland).

NCRP (1972). National Council on Radiation Protection and Measurements. *Protection Against Radiation from Brachytherapy Sources*, NCRP Report No. 40 (National Council on Radiation Protection and Measurements, Bethesda, Maryland).

NCRP (1976a). National Council on Radiation Protection and Measurements. *Structural Shielding Design and Evaluation for Medical Use of X-Rays and Gamma Rays of Energies up to 10 MeV*, NCRP Report No. 49 (National Council on Radiation Protection and Measurements, Bethesda, Maryland).

NCRP (1976b). National Council on Radiation Protection and Measurements. *Tritium Measurement Techniques*, NCRP Report No. 47 (National Council on Radiation Protection and Measurements, Bethesda, Maryland).

NCRP (1976c). National Council on Radiation Protection and Measurements *Radiation Protection for Medical and Allied Health Personnel*, NCRP Report No. 48 (National Council on Radiation Protection and Measurements, Bethesda, Maryland) Superseded by this Report.

NCRP (1977a). National Council on Radiation Protection and Measurements. *Review of NCRP Radiation Dose Limit for Embryo and Fetus in*

Occupationally Exposed Women, NCRP Report No. 53 (National Council on Radiation Protection and Measurements, Bethesda, Maryland).

NCRP (1977b). National Council on Radiation Protection and Measurements. *Radiation Protection Design Guidelines for 0.1 - 100 MeV Particle Accelerator Facilities*, NCRP Report No. 51 (National Council on Radiation Protection and Measurements, Bethesda, Maryland).

NCRP (1978a). National Council on Radiation Protection and Measurements. *Operational Radiation Safety Program*, NCRP Report No. 59 (National Council on Radiation Protection and Measurements, Bethesda, Maryland).

NCRP (1978b). National Council on Radiation Protection and Measurements. *Instrumentation and Monitoring Methods for Radiation Protection*, NCRP Report No. 57 (National Council on Radiation Protection and Measurements, Bethesda, Maryland).

NCRP (1980a). National Council on Radiation Protection and Measurements. *Influence of Dose and Its Distribution in Time on Dose-Response Relationships for Low-LET Radiations*, NCRP Report No. 64 (National Council on Radiation Protection and Measurements, Bethesda, Maryland).

NCRP (1980b). National Council on Radiation Protection and Measurements. *Management of Persons Accidentally Contaminated with Radionuclides*, NCRP Report No. 65 (National Council on Radiation Protection and Measurements, Bethesda, Maryland).

NCRP (1981). National Council on Radiation Protection and Measurements. *Radiation Protection in Pediatric Radiology*, NCRP Report No. 68 (National Council on Radiation Protection and Measurements, Bethesda, Maryland).

NCRP (1982). National Council on Radiation Protection and Measurements. *Nuclear Medicine - Factors Influencing the Choice and Use of Radionuclides in Diagnosis and Therapy*, NCRP Report No. 70 (National Council on Radiation Protection and Measurements, Bethesda, Maryland).

NCRP (1983a). National Council on Radiation Protection and Measurements. *Operational Radiation Safety-Training*, NCRP Report No. 71 (National Council on Radiation Protection and Measurements, Bethesda, Maryland).

NCRP (1983b). National Council on Radiation Protection and Measurements. *Biological Effects of Ultrasound: Mechanisms and Clinical Implications*, NCRP Report No. 74 (National Council on Radiation Protection and Measurements, Bethesda, Maryland).

NCRP (1983c). National Council on Radiation Protection and Measurements. *Protection in Nuclear Medicine and Ultrasound Diagnostic Procedures in Children*, NCRP Report No. 73 (National Council on Radiation Protection and Measurements, Bethesda, Maryland).

NCRP (1984). National Council on Radiation Protection and Measurements. *Neutron Contamination from Medical Electron Accelerators*, NCRP Report No. 79 (National Council on Radiation Protection and Measurements, Bethesda, Maryland)

NCRP (1985a). National Council on Radiation Protection and Measurements. *SI Units in Radiation Protection and Measurements*, NCRP Report No. 82 (National Council on Radiation Protection and Measurements, Bethesda, Maryland).

NCRP (1985b). National Council on Radiation Protection and Measurements. *A Handbook of Radioactivity Measurements Procedures 2nd ed.*, NCRP Report No. 58 (National Council on Radiation Protection and Measurements, Bethesda, Maryland).

NCRP (1986). National Council on Radiation Protection and Measurements. *Biological Effects and Exposure Criteria for Radiofrequency Electromagnetic Fields*, NCRP Report No. 86 (National Council on Radiation Protection and Measurements, Bethesda, Maryland).

NCRP (1987a). National Council on Radiation Protection and Measurements. *Recommendations on Limits for Exposure to Ionizing Radiation*, NCRP Report No. 91 (National Council on Radiation Protection and Measurements, Bethesda, Maryland).

NCRP (1987b). National Council on Radiation Protection and Measurements. *Use of Bioassay Procedures for Assessment of Internal Radionuclide Deposition*, NCRP Report No. 87 (National Council on Radiation Protection and Measurements, Bethesda, Maryland).

NCRP (1987c) National Council on Radiation Protection and Measurements. *Radiation Alarms and Access Control Systems*, NCRP Report No. 88 (National Council on Radiation Protection and Measurements, Bethesda, Maryland).

NCRP (1987d). National Council on Radiation Protection and Measurements. *Ionizing Radiation Exposure of the Population of the United States*, NCRP Report No. 93 (National Council on Radiation Protection and Measurements, Bethesda, Maryland).

NCRP (1988a). National Council on Radiation Protection and Measurements. *Exposure to the Population in the United States and Canada from Natural Background Radiation*, NCRP Report No. 94 (National Council on Radiation Protection and Measurements, Bethesda, Maryland).

NCRP (1988b). National Council on Radiation Protection and Measurements. *Radiation Exposure of the U.S. Population from Consumer Products and Miscellaneous Sources*, NCRP Report No. 95 (National Council on Radiation Protection and Measurements, Bethesda, Maryland).

NCRP (1989a). National Council on Radiation Protection and Measurements. *Exposure of the U.S. Population from Diagnostic Medical Radiation,* NCRP Report No. 100 (National Council on Radiation Protection and Measurements, Bethesda, Maryland).

NCRP (1989b). National Council on Radiation Protection and Measurements. *Exposure of the U.S. Population from Occupational Radiation,* NCRP Report No. 101, (National Council on Radiation Protection and Measurements, Bethesda, Maryland).

NRC (1983). National Research Council. *Video Displays, Work and Vision* (National Academy Press, Washington).

NRC (1987). Nuclear Regulatory Commission, 10CFR 35.75 (Government Printing Office Washington).

OTAKE, M. AND SCHULL, W.J. (1984) *"In utero* exposure to A-bomb radiation and mental retardation; a reassessment," Br. J. Radiol. *57,* 409-414.

SCHONKEN, P. MARCHAL, G., COENEN, Y., BAERT, A.L. AND PONETTE, E. (1978). "Body and gonad dose in computer tomography of the trunk," J. Belge Radiol. **61** (4), 363-371.

SEER (1981). *Surveillance, Epidemiology, and End Results: Incidence and Mortality Data, 1973-77.* National Cancer Institute Monograph 57 (Government Printing Office, Washington).

SHOPE, T.B., MORGAN, T.J., SHOWALTER, C.K., PENTLOW, K.S., ROTHENBERG, L.N, WHITE, D.R. AND SPELLER, R.D. (1982). "Radiation dosimetry survey of computed tomography systems from ten manufacturers," Br. J. Radiol. **55,** 60-69.

SYED, I.B., FLOWERS, N., GRANLICK, D. AND SAMOLS, E. (1982). "Radiation exposures in nuclear cardiovascular studies," Health Phys. **42,** 159-163.

UNSCEAR (1986). United Nations Scientific Committee on the Effects of Atomic Radiation. *Genetic and Somatic Effects of Ionizing Radiation,* United Nations Scientific Committee on the Effects of Atomic Radiation ,1986 Report to the General Assembly, with annexes (United Nations, New York).

UNSCEAR (1988). United Nations Scientific Committee on the Effects of Atomic Radiation. *Sources, Effects and Risks of Ionizing Radiation,* UNSCEAR 1988 Report to the General Assembly with Annexes (United Nations, New York.)

ZIEMER, P.L. AND ORVIS, A.L. (1981) "Hospital Radiation Safety" page 275 in *Handbook of Hospital Safety,* Stanley, P.E., Ed. (CRC Press, Boca Raton, Florida).

The NCRP

The National Council on Radiation Protection and Measurements is a nonprofit corporation chartered by Congress in 1964 to:

1. Collect, analyze, develop, and disseminate in the public interest information and recommendations about (a) protection against radiation and (b) radiation measurements, quantities, and units, particularly those concerned with radiation protection;
2. Provide a means by which organizations concerned with the scientific and related aspects of radiation protection and of radiation quantities, units, and measurements may cooperate for effective utilization of their combined resources, and to stimulate the work of such organizations;
3. Develop basic concepts about radiation quantities, units, and measurements, about the application of these concepts, and about radiation protection;
4. Cooperate with the International Commission on Radiological Protection, the International Commission on Radiation Units and Measurements, and other national and international organizations, governmental and private, concerned with radiation quantities, units, and measurements and with radiation protection.

The Council is the successor to the unincorporated association of scientists known as the National Committee on Radiation Protection and Measurements and was formed to carry on the work begun by the Committee.

The Council is made up of the members and the participants who serve on the over sixty scientific committees of the Council. The scientific committees, composed of experts having detailed knowledge and competence in the particular area of the committee's interest draft proposed recommendations. These are then submitted to the full membership of the Council for careful review and approval before being published.

The following comprise the current officers and membership of the Council:

108

Members

SEYMOUR ABRAHAMSON	ETHEL S. GILBERT	A. ALAN MOGHISSI
S. JAMES ADELSTEIN	ROBERT A. GOEPP	MARY ELLEN O'CONNOR
PETER R. ALMOND	JOEL E. GRAY	ANDREW K. POZNANSKI
EDWARD L. ALPEN	ARTHUR W. GUY	NORMAN C. RASMUSSEN
LYNN R. ANSPAUGH	ERIC J. HALL	CHESTER R. RICHMOND
JOHN A. AUXIER	NAOMI H. HARLEY	MARVIN ROSENSTEIN
WILLIAM J. BAIR	WILLIAM R. HENDEE	LAWRENCE N. ROTHENBERG
MICHAEL A. BENDER	DONALD G. JACOBS	LEONARD A. SAGAN
BRUCE B. BOECKER	A. EVERETTE JAMES, JR.	KEITH J. SCHIAGER
JOHN D. BOICE, JR.	BERND KAHN	ROBERT A. SCHLENKER
ROBERT L. BRENT	KENNETH R. KASE	WILLIAM J. SCHULL
ANTONE BROOKS	CHARLES E. LAND	ROY E. SHORE
MELVIN W. CARTER	GEORGE R. LEOPOLD	WARREN K. SINCLAIR
RANDALL S. CASWELL	RAY D. LLOYD	PAUL SLOVIC
JAMES E. CLEAVER	HARRY R. MAXON	RICHARD A. TELL
FRED T. CROSS	ROGER O. MCCLELLAN	WILLIAM L. TEMPLETON
STANLEY B. CURTIS	JAMES E. MCLAUGHLIN	THOMAS S. TENFORDE
GERALD D. DODD	BARBARA J. MCNEIL	J. W. THIESSEN
PATRICIA W. DURBIN	THOMAS F. MEANEY	JOHN E. TILL
CHARLES EISENHAUER	CHARLES B. MEINHOLD	ROBERT ULLRICH
THOMAS S. ELY	MORTIMER L. MENDELSOHN	ARTHUR C. UPTON
JACOB I. FABRIKANT	FRED A. METTLER	GEORGE L. VOELZ
R. J. MICHAEL FRY	WILLIAM A. MILLS	GEORGE M. WILKENING
THOMAS F. GESELL	DADE W. MOELLER	MARVIN ZISKIN

Honorary Members

LAURISTON S. TAYLOR, *Honorary President*

VICTOR P. BOND	ROBERT O. GORSON	WESLEY L. NYBORG
REYNOLD F. BROWN	JOHN H. HARLEY	HARALD H. ROSSI
AUSTIN M. BRUES	JOHN W. HEALY	WILLIAM L. RUSSELL
GEORGE W. CASARETT	LOUIS H.	JOHN H. RUST
FREDERICK P. COWAN	HEMPELMANN, JR.	EUGENE L. SAENGER
JAMES F. CROW	PAUL C. HODGES	J. NEWELL STANNARD
MERRILL EISENBUD	GEORGE V. LEROY	JOHN B. STORER
ROBLEY D. EVANS	WILFRID B. MANN	ROY C. THOMPSON
RICHARD F. FOSTER	KARL Z. MORGAN	EDWARD W. WEBSTER
HYMER L. FRIEDELL	ROBERT J. NELSEN	HAROLD O. WYCKOFF

Currently, the following subgroups are actively engaged in formulating recommendations:

SC 1:   Basic Radiation Protection Criteria
         SC 1-1 Probability of Causation for Genetic and Developmental
               Effects
         SC 1-2 The Assessment of Risk for Radiation Protection Purposes
SC 16:   X-Ray Protection in Dental Offices
SC 40:   Biological Aspects of Radiation Protection Criteria

SC 40-1 Atomic Bomb Survivor Dosimetry
SC 45:   Radiation Received by Radiation Employees
SC 46:   Operational Radiation Safety
SC 46-2 Uranium Mining and Milling—Radiation Safety Programs
SC 46-3 ALARA for Occupationally Exposed Individuals in Clinical
          Radiology
SC 46-4 Calibration of Survey Instrumentation
SC 46-5 Maintaining Radiation Protection Records
SC 46-7 Emergency Planning
SC 46-8 Radiation Protection Design Guidelines for Particle
          Accelerator Facilities
SC 46-9 ALARA at Nuclear Plants
SC 46-10 Assessment of Occupational Doses from Internal Emitters
SC 46-11 Radiation Protection During Special Medical Procedures
SC 52:   Conceptual Basis of Calculations of Dose Distributions
SC 57:   Internal Emitter Standards
SC 57-2 Respiratory Tract Model
SC 57-6 Bone Problems
SC 57-8 Leukemia Risk
SC 57-9 Lung Cancer Risk
SC 57-10 Liver Cancer Risk
SC 57-12 Strontium
SC 57-14 Placental Transfer
SC 57-15 Uranium
SC 59:   Human Population Exposure Experience
SC 63:   Radiation Exposure Control in a Nuclear Emergency
SC 63-1 Public Knowledge About Radiation
SC 63-2 Criteria for Radiation Instruments for the Public
SC 64:   Environmental Radioactivity and Waste Management
SC 64-6 Screening Models
SC 64-7 Contaminated Soil as a Source of Radiation Exposure
SC 64-8 Ocean Disposal of Radioactive Waste
SC 64-9 Effects of Radiation on Aquatic Organisms
SC 64-10 Xenon
SC 64-11 Disposal of Low Level Waste
SC 65:   Quality Assurance and Accuracy in Radiation Protection
          Measurements
SC 66:   Biological Effects and Exposure Criteria for Ultrasound
SC 67:   Biological Effects of Magnetic Fields
SC 68:   Microprocessors in Dosimetry
SC 69:   Efficacy of Radiographic Procedures
SC 71:   Radiation Exposure and Potentially Related Injury
SC 74:   Radiation Received in the Decontamination of Nuclear Facilities
SC 76:   Effects of Radiation on the Embryo-Fetus
SC 77:   Guidance on Occupational and Public Exposure Resulting from
          Diagnostic Nuclear Medicine Procedures
SC 78:   Practical Guidance on the Evaluation of Human Exposures to
          Radiofrequency Radiation
SC 79:   Extremely Low-Frequency Electric and Magnetic Fields
SC 80:   Radiation Biology of the Skin (Beta-Ray Dosimetry)
SC 80-1 Hot Particles on the Skin
SC 81:   Assessment of Exposures from Therapy

SC 82:   Control of Indoor Radon
SC 83:   Identification of Research Needs

Study Group on Comparative Risk
Ad Hoc Group on Video Display Terminals
Task Force on Occupational Exposure Levels

In recognition of its responsibility to facilitate and stimulate cooperation among organizations concerned with the scientific and related aspects of radiation protection and measurement, the Council has created a category of NCRP Collaborating Organizations. Organizations or groups of organizations that are national or international in scope and are concerned with scientific problems involving radiation quantities, units, measurements, and effects, or radiation protection may be admitted to collaborating status by the Council. The present Collaborating Organizations with which the NCRP maintains liaison are as follows:

American Academy of Dermatology
American Association of Physicists in Medicine
American College of Medical Physics
American College of Nuclear Physicians
American College of Radiology
American Dental Association
American Industrial Hygiene Association
American Institute of Ultrasound in Medicine
American Insurance Services Group
American Medical Association
American Nuclear Society
American Occupational Medical Association
American Podiatric Medical Association
American Public Health Association
American Radium Society
American Roentgen Ray Society
American Society of Radiologic Technologists
American Society for Therapeutic Radiology and Oncology
Association of University Radiologists
Bioelectromagnetics Society
College of American Pathologists
Conference of Radiation Control Program Directors
Electric Power Research Institute
Federal Communications Commission
Federal Emergency Management Agency
Genetics Society of America
Health Physics Society
Institute of Nuclear Power Operations
National Electrical Manufacturers Association
National Institute of Standards and Technology
Nuclear Management and Resources Council
Radiation Research Society
Radiological Society of North America

Society of Nuclear Medicine
United States Air Force
United States Army
United States Department of Energy
United States Department of Housing and Urban Development
United States Department of Labor
United States Environmental Protection Agency
United States Navy
United States Nuclear Regulatory Commission
United States Public Health Service

The NCRP has found its relationships with these organizations to be extremely valuable to continued progress in its program.

Another aspect of the cooperative efforts of the NCRP relates to the special liaison relationships established with various governmental organizations that have an interest in radiation protection and measurements. This liaison relationship provides: (1) an opportunity for participating organizations to designate an individual to provide liaison between the organization and the NCRP; (2) that the individual designated will receive copies of draft NCRP reports (at the time that these are submitted to the members of the Council) with an invitation to comment, but not vote; and (3) that new NCRP efforts might be discussed with liaison individuals as appropriate, so that they might have an opportunity to make suggestions on new studies and related matters. The following organizations participate in the special liaison program:

Australian Radiation Laboratory
Commissariat a l'Energie Atomique (France)
Commission of the European Communities
Defense Nuclear Agency
Federal Emergency Management Agency
Japan Radiation Council
National Institute of Standards and Technology
National Radiological Protection Board (United Kingdom)
National Research Council (Canada)
Office of Science and Technology Policy
Office of Technology Assessment
United States Air Force
United States Army
United States Coast Guard
United States Department of Energy
United States Department of Health and Human Services
United States Department of Labor
United States Department of Transportation
United States Environmental Protection Agency
United States Navy
United States Nuclear Regulatory Commission

The NCRP values highly the participation of these organizations in the liaison program.

The Council's activities are made possible by the voluntary contribution of time and effort by its members and participants and the generous support of the following organizations:

Alfred P. Sloan Foundation
Alliance of American Insurers
American Academy of Dental Radiology
American Academy of Dermatology
American Association of Physicists in Medicine
American College of Nuclear Physicians
American College of Radiology
American College of Radiology Foundation
American Dental Association
American Hospital Radiology Administrators
American Industrial Hygiene Association
American Insurance Services Group
American Medical Association
American Nuclear Society
American Occupational Medical Association
American Osteopathic College of Radiology
American Podiatric Medical Association
American Public Health Association
American Radium Society
American Roentgen Ray Society
American Society of Radiologic Technologists
American Society for Therapeutic Radiology and Oncology
American Veterinary Medical Association
American Veterinary Radiology Society
Association of University Radiologists
Atomic Industrial Forum
Battelle Memorial Institute
Center for Devices and Radiological Health
College of American Pathologists
Commonwealth of Pennsylvania
Defense Nuclear Agency
Edison Electric Institute
Edward Mallinckrodt, Jr. Foundation
EG&G Idaho, Inc.
Electric Power Research Institute
Federal Emergency Management Agency
Florida Institute of Phosphate Research
Genetics Society of America
Health Physics Society
Institute of Nuclear Power Operations
James Picker Foundation
Richard Lounsbery Foundation
National Aeronautics and Space Administration
National Association of Photographic Manufacturers
National Bureau of Standards
National Cancer Institute

National Electrical Manufacturers Association
Nuclear Management and Resources Council
Radiation Research Society
Radiological Society of North America
Sandia National Laboratory
Society of Nuclear Medicine
United States Department of Energy
United States Department of Labor
United States Environmental Protection Agency
United States Navy
United States Nuclear Regulatory Commission

To all of these organizations the Council expresses its profound appreciation for their support.

Initial funds for publication of NCRP reports were provided by a grant from the James Picker Foundation and for this the Council wishes to express its deep appreciation.

The NCRP seeks to promulgate information and recommendations based on leading scientific judgment on matters of radiation protection and measurement and to foster cooperation among organizations concerned with these matters. These efforts are intended to serve the public interest and the Council welcomes comments and suggestions on its reports or activities from those interested in its work.

NCRP Publications

NCRP publications are distributed by the NCRP Publications' office. Information on prices and how to order may be obtained by directing an inquiry to:

> NCRP Publications
> 7910 Woodmont Ave., Suite 800
> Bethesda, Md 20814

The currently available publications are listed below.

Proceedings of the Annual Meeting

No.	Title
1	*Perceptions of Risk,* Proceedings of the Fifteenth Annual Meeting, Held on March 14–15, 1979 (Including Taylor Lecture No. 3) (1980)
2	*Quantitative Risk in Standards Setting,* Proceedings of the Sixteenth Annual Meeting, Held on April 2–3, 1980 (Including Taylor Lecture No. 4) (1981)
3	*Critical Issues in Setting Radiation Dose Limits,* Proceedings of the Seventeenth Annual Meeting, Held on April 8–9, 1981 (Including Taylor Lecture No. 5) (1982)
4	*Radiation Protection and New Medical Diagnostic Procedures,* Proceedings of the Eighteenth Annual Meeting, Held on April 6–7, 1982 (Including Taylor Lecture No. 6) (1983)
5	*Environmental Radioactivity,* Proceedings of the Nineteenth Annual Meeting, Held on April 6–7, 1983 (Including Taylor Lecture No. 7) (1984)
6	*Some Issues Important in Developing Basic Radiation Protection Recommendations,* Proceedings of the Twentieth Annual Meeting, Held on April 4–5, 1984 (Including Taylor Lecture No. 8) (1985)
7	*Radioactive Waste,* Proceedings of the Twenty-first Annual Meeting, Held on April 3–4, 1985 (Including Taylor Lecture No. 9) (1986)

8   *Nonionizing Electromagnetic Radiation and Ultrasound,* Proceedings of the Twenty-second Annual Meeting, Held on April 2–3, 1986 (Including Taylor Lecture No. 10) (1988)

9   *New Dosimetry at Hiroshima and Nagasaki and Its Implications for Risk Estimates,* Proceedings of the Twenty-third Annual Meeting, Held on April 5–6, 1987 (Including Taylor Lecture No. 11) (1988).

10   *Radon,* Proceedings of the Twenty-fourth Annual Meeting, Held on March 30-31, 1988 (Including Taylor Lecture No. 12) (1989).

Symposium Proceedings

The Control of Exposure of the Public to Ionizing Radiation in the Event of Accident or Attack, Proceedings of a Symposium held April 27–29, 1981 (1982)

Lauriston S. Taylor Lectures

No.   Title and Author

1   *The Squares of the Natural Numbers in Radiation Protection* by Herbert M. Parker (1977)

2   *Why be Quantitative About Radiation Risk Estimates?* by Sir Edward Pochin (1978)

3   *Radiation Protection—Concepts and Trade Offs* by Hymer L. Friedell (1979) [Available also in *Perceptions of Risk,* see above]

4   *From "Quantity of Radiation" and "Dose" to "Exposure" and "Absorbed Dose"—An Historical Review* by Harold O. Wyckoff (1980) [Available also in *Quantitative Risks in Standards Setting,* see above]

5   *How Well Can We Assess Genetic Risk? Not Very* by James F. Crow (1981) [Available also in *Critical Issues in Setting Radiation Dose Limits,* see above]

6   *Ethics, Trade-offs and Medical Radiation* by Eugene L. Saenger (1982) [Available also in *Radiation Protection and New Medical Diagnostic Approaches,* see above]

7   *The Human Environment-Past, Present and Future* by Merril Eisenbud (1983) [Available also in *Environmental Radioactivity,* see above]

8   *Limitation and Assessment in Radiation Protection* by
     Harald H. Rossi (1984) [Available also in *Some Issues
     Important in Developing Basic Radiation Protection
     Recommendations,* see above]

9   *Truth (and Beauty) in Radiation Measurement* by John
     H. Harley (1985) [Available also in *Radioactive Waste,*
     see above]

10   *Nonionizing Radiation Bioeffects: Cellular Properties and
     Interactions* by Herman P. Schwan (1986) [Available
     also in *Nonionizing Electromagnetic Radiations and
     Ultrasound,* see above]

11   *How to be Quantitative about Radiation Risk Estimates*
     by Seymour Jablon (1987) [Available also in *New Dosi-
     metry at Hiroshima and Nagasaki and its Implications
     for Risk Estimates,* see above]

12   *How Safe is Safe Enough?* by Bo Lindell (1988) [Avail-
     able also in *Radon,* See above]

13   *Radiobiology and Radiation Protection: The Past Cen-
     tury and Prospects for the Future* by Arthur C. Upton
     (1989)

NCRP Commentaries

No.   Title

1   *Krypton-85 in the Atmosphere—With Specific Reference
     to the Public Health Significance of the Proposed Con-
     trolled Release at Three Mile Island* (1980)

2   *Preliminary Evaluation of Criteria for the Disposal of
     Transuranic Contaminated Waste* (1982)

3   *Screening Techniques for Determining Compliance with
     Environmental Standards* (1986), Rev. (1989)

4   *Guidelines for the Release of Waste Water from Nuclear
     Facilities with Special Reference to the Public Health
     Significance of the Proposed Release of Treated Waste
     Waters at Three Mile Island* (1987)

5   *Living Without Landfills* (1989)

NCRP Reports

No.   Title

8   *Control and Removal of Radioactive Contamination in
     Laboratories* (1951)

22   *Maximum Permissible Body Burdens and Maximum Permissible Concentrations of Radionuclides in Air and in Water for Occupational Exposure* (1959) [Includes Addendum 1 issued in August 1963]

23   *Measurement of Neutron Flux and Spectra for Physical and Biological Applications* (1960)

25   *Measurement of Absorbed Dose of Neutrons and Mixtures of Neutrons and Gamma Rays* (1961)

27   *Stopping Powers for Use with Cavity Chambers* (1961)

30   *Safe Handling of Radioactive Materials* (1964)

32   *Radiation Protection in Educational Institutions* (1966)

35   *Dental X-Ray Protection* (1970)

36   *Radiation Protection in Veterinary Medicine* (1970)

37   *Precautions in the Management of Patients Who Have Received Therapeutic Amounts of Radionuclides* (1970)

38   *Protection Against Neutron Radiation* (1971)

40   *Protection Against Radiation from Brachytherapy Sources* (1972)

41   *Specifications of Gamma-Ray Brachytherapy Sources* (1974)

42   *Radiological Factors Affecting Decision-Making in a Nuclear Attack* (1974)

44   *Krypton-85 in the Atmosphere—Accumulation, Biological Significance, and Control Technology* (1975)

46   *Alpha-Emitting Particles in Lungs* (1975)

47   *Tritium Measurement Techniques* (1976)

49   *Structural Shielding Design and Evaluation for Medical Use of X Rays and Gamma Rays of Energies Up to 10 MeV* (1976)

50   *Environmental Radiation Measurements* (1976)

51   *Radiation Protection Design Guidelines for 0.1–100 MeV Particle Accelerator Facilities* (1977)

52   *Cesium-137 from the Environment to Man: Metabolism and Dose* (1977)

53   *Review of NCRP Radiation Dose Limit for Embryo and Fetus in Occupationally Exposed Women* (1977)

54   *Medical Radiation Exposure of Pregnant and Potentially Pregnant Women* (1977)

55   *Protection of the Thyroid Gland in the Event of Releases of Radioiodine* (1977)

57   *Instrumentation and Monitoring Methods for Radiation Protection* (1978)

58   *A Handbook of Radioactivity Measurements Procedures, 2nd ed.* (1985)

Binders for NCRP Reports are available. Two sizes make it possible to collect into small binders the "old series" of reports (NCRP Reports Nos. 8–30) and into large binders the more recent publications (NCRP Reports Nos. 32–105). Each binder will accommodate from five to seven reports. The binders carry the identification "NCRP Reports" and come with label holders which permit the user to attach labels showing the reports contained in each binder.

The following bound sets of NCRP Reports are also available:

Volume I. NCRP Reports Nos. 8, 22
Volume II. NCRP Reports Nos. 23, 25, 27, 30
Volume III. NCRP Reports Nos. 32, 35, 36, 37
Volume IV. NCRP Reports Nos. 38, 40, 41
Volume V. NCRP Reports Nos. 42, 44, 46
Volume VI. NCRP Reports Nos. 47, 49, 50, 51
Volume VII. NCRP Reports Nos. 52, 53, 54, 55, 57
Volume VIII. NCRP Reports No. 58
Volume IX. NCRP Reports Nos. 59, 60, 61, 62, 63
Volume X. NCRP Reports Nos. 64, 65, 66, 67
Volume XI. NCRP Reports Nos. 68, 69, 70, 71, 72
Volume XII. NCRP Reports Nos. 73, 74, 75, 76
Volume XIII. NCRP Reports Nos. 77, 78, 79, 80
Volume XIV. NCRP Reports Nos. 81, 82, 83, 84, 85
Volume XV. NCRP Reports Nos. 86, 87, 88, 89
Volume XVI. NCRP Reports Nos. 90, 91, 92, 93
Volume XVII. NCRP Reports Nos. 94, 95, 96, 97

(Titles of the individual reports contained in each volume are given above).

The following NCRP Reports are now superseded and/or out of print:

No.	Title
1	*X-Ray Protection* (1931). [Superseded by NCRP Report No. 3]
2	*Radium Protection* (1934). [Superseded by NCRP Report No. 4]
3	*X-Ray Protection* (1936). [Superseded by NCRP Report No. 6]
4	*Radium Protection* (1938). [Superseded by NCRP Report No. 13]
5	*Safe Handling of Radioactive Luminous Compounds* (1941). [Out of Print]
6	*Medical X-Ray Protection Up to Two Million Volts* (1949). [Superseded by NCRP Report No. 18]
7	*Safe Handling of Radioactive Isotopes* (1949). [Superseded by NCRP Report No. 30]

9 *Recommendations for Waste Disposal of Phosphorus-32 and Iodine-131 for Medical Users* (1951). [Out of Print]

10 *Radiological Monitoring Methods and Instruments* (1952). [Superseded by NCRP Report No. 57]

11 *Maximum Permissible Amounts of Radioisotopes in the Human Body and Maximum Permissible Concentrations in Air and Water* (1953). [Superseded by NCRP Report No. 22]

12 *Recommendations for the Disposal of Carbon-14 Wastes* (1953). [Superseded by NCRP Report No. 81]

13 *Protection Against Radiations from Radium, Cobalt-60 and Cesium-137* (1954). [Superseded by NCRP Report No. 24]

14 *Protection Against Betatron—Synchrotron Radiations Up to 100 Million Electron Volts* (1954). [Superseded by NCRP Report No. 51]

15 *Safe Handling of Cadavers Containing Radioactive Isotopes* (1953). [Superseded by NCRP Report No. 21]

16 *Radioactive Waster Disposal in the Ocean* (1954). [Out of Print]

17 *Permissible Dose from External Sources of Ionizing Radiation* (1954) including *Maximum Permissible Exposure to Man, Addendum to National Bureau of Standards Handbook 59* (1958). [Superseded by NCRP Report No. 39]

18 *X-Ray Protection* (1955). [Superseded by NCRP Report No. 26]

19 *Regulation of Radiation Exposure by Legislative Means* (1955). [Out of Print]

20 *Protection Against Neutron Radiation Up to 30 Million Electron Volts* (1957). [Superseded by NCRP Report No. 38]

21 *Safe Handling of Bodies Containing Radioactive Isotopes* (1958). [Superseded by NCRP Report No. 37]

24 *Protection Against Radiations from Sealed Gamma Sources* (1960). [Superseded by NCRP Report Nos. 33, 34, and 40]

26 *Medical X-Ray Protection Up to Three Million Volts* (1961). [Superseded by NCRP Report Nos. 33, 34, 35, and 36]

28 *A Manual of Radioactivity Procedures* (1961). [Superseded by NCRP Report No. 58]

29 *Exposure to Radiation in an Emergency* (1962). [Superseded by NCRP Report No. 42]

31   *Shielding for High Energy Electron Accelerator Installations* (1964). [Superseded by NCRP Report No. 51]

33   *Medical X-Ray and Gamma-Ray Protection for Energies up to 10 MeV*—Equipment Design and Use (1968). [Superseded by NCRP Report No. 102]

34   *Medical X-Ray and Gamma-Ray Protection for Energies Up to 10 MeV—Structural Shielding Design and Evaluation* (1970). [Superseded by NCRP Report No. 49]

39   *Basic Radiation Protection Criteria* (1971). [Superseded by NCRP Report No. 91]

43   *Review of the Current State of Radiation Protection Philosophy* (1975). [Superseded by NCRP Report No. 91]

45   *Natural Background Radiation in the United States* (1975). [Superseded by NCRP Report No. 94]

48   *Radiation Protection for Medical and Allied Health Personnel* [Superseded by NCRP Report No. 105]

56   *Radiation Exposure from Consumer Products and Miscellaneous Sources* (1977). [Superseded by NCRP Report No. 95]

58   *A Handbook on Radioactivity Measurement Procedures.* [Superseded by NCRP Report No. 58, 2nd ed.]

Other Documents

The following documents of the NCRP were published outside of the NCRP Reports and Commentaries series:

"Blood Counts, Statement of the National Committee on Radiation Protection," Radiology 63, 428 (1954)

"Statements on Maximum Permissible Dose from Television Receivers and Maximum Permissible Dose to the Skin of the Whole Body," Am. J. Roentgenol., Radium Ther. and Nucl. Med. 84, 152 (1960) and Radiology 75, 122 (1960)

Dose Effect Modifying Factors In Radiation Protection, Report of Subcommittee M-4 (Relative Biological Effectiveness) of the National Council on Radiation Protection and Measurements, Report BNL 50073 (T-471) (1967) Brookhaven National Laboratory (National Technical Information Service, Springfield, Virginia).

X-Ray Protection Standards for Home Television Receivers, Interim Statement of the National Council on Radiation Protection and Measurements (National Council on Radiation Protection and Measurements, Washington, 1968)

Specification of Units of Natural Uranium and Natural Thorium (National Council on Radiation Protection and Measurements, Washington, 1973)

NCRP Statement on Dose Limit for Neutrons (National Council on Radiation Protection and Measurements, Washington, 1980)

Control of Air Emissions of Radionuclides (National Council on Radiation Protection and Measurements, Bethesda, Maryland, 1984)

Copies of the statements published in journals may be consulted in libraries. A limited number of copies of the remaining documents listed above are available for distribution by NCRP Publications.

Index